More information about this series at https://link.springer.com/bookseries/7899

Communications
in Computer and Information Science

1630

Filippo Neri · Ke-Lin Du ·
Vijaya Kumar Varadarajan ·
San-Blas Angel-Antonio · Zhiyu Jiang (Eds.)

Computer and Communication Engineering

2nd International Conference, CCCE 2022
Rome, Italy, March 11–13, 2022
Revised Selected Papers

Springer

Editors
Filippo Neri (ID)
University of Naples Federico II
Naples, Italy

Ke-Lin Du (ID)
Concordia University
Montreal, QC, Canada

Vijaya Kumar Varadarajan (ID)
The University of New South Wales
Sydney, NSW, Australia

San-Blas Angel-Antonio (ID)
Miguel Hernandez University
Elche, Spain

Zhiyu Jiang (ID)
Northwestern Polytechnical University
Xi'an, China

ISSN 1865-0929   ISSN 1865-0937 (electronic)
Communications in Computer and Information Science
ISBN 978-3-031-17421-6   ISBN 978-3-031-17422-3 (eBook)
https://doi.org/10.1007/978-3-031-17422-3

This Springer imprint is published by the registered company Springer Nature Switzerland AG
The registered company address is: Gewerbestrasse 11, 6330 Cham, Switzerland

Preface

It is my pleasure to present you the proceedings of the 2nd International Conference on Computer and Communication Engineering (CCCE 2022), which was held virtually via Rome, Italy, during March 11–13, 2022.

CCCE provides a scientific platform for both local and international scientists, researchers, engineers, and technologists who work in all areas of computer and communication engineering to share not just technical interests but, among other things, culture and history. This year, we received 36 submissions, of which 17 were selected for presentation and publication (an acceptance rate of 47.2%). Each submission was reviewed by at least 2 members of the Technical Program Committee in a double-blind manner.

In addition to the contributed papers, internationally known experts from several countries were also invited to deliver keynote and invited speeches at CCCE 2022. We are grateful to Angrisani Leopoldo (Università degli Studi di Napoli Federico II, Italy), Assaf Schuster (Israel Institute of Technology, Israel), Ning Xiong (Mälardalen University, Sweden), and Dapeng Wu (University of Florida, USA) for their insightful talks. The whole conference was held online in Zoom, and we also had tests for each presenter in advance to ensure the successful delivery of the conference. For those who had internet problems, a pre-recorded video presentation was accepted as an alternative. Meanwhile, the whole conference was recorded as a conference backup only.

I would like to take this opportunity to thank all the participants of CCCE 2022 for sharing their latest knowledge and findings on a variety of topics, including 5th generation networks (5G), 5G core networks and service-based architecture, artificial intelligence (AI), and computer networks and mobile communication.

On behalf of the organizing committee, we hope that you found this conference to be a fruitful meeting place and enjoy reading the proceedings. We will continue to organize this conference in the future to provide an effective platform for further exchange of new knowledge and perhaps potential collaboration in the research areas of computer and communication engineering.

August 2022

Filippo Neri
Ning Xiong
Takao Terano
Ke-Lin Du
Vijayakumar Varadarajan
Angel-Antonio San-Blas
Zhiyu Jiang

Organization

Advisory Committee

Angrisani Leopoldo   Università degli Studi di Napoli Federico II, Italy
Assaf Schuster   Israel Institute of Technology, Israel
Dapeng Wu   University of Florida, USA

Conference Chair

Filippo Neri   University of Naples "Federico II", Italy

Program Chairs

Ning Xiong   Mälardalen University, Sweden
Takao Terano   Chiba University of Commerce, Japan
Ke-Lin Du   Concordia University, Canada
Vijayakumar Varadarajan   University of New South Wales, Australia

Program Co-chairs

Angel-Antonio San-Blas   Miguel Hernández University of Elche, Spain
Zhiyu Jiang   Northwestern Polytechnical University, China

Technical Program Committee

Qian Han   Dartmouth College, USA
Pavel Kromer   VSB-Technical University of Ostrava, Czech Republic
Nazli Siasi   Newport University, USA
Emanuela Marasco   George Mason University, USA
Avksentieva Elena Yurevna   ITMO University, Russia
Nitikarn Nimsuk   Thammasat University, Thailand
Mahmoud M. Elmesalawy   Helwan University, Egypt
Kadhim Hayawi   Zayed University, UAE
Toufik Bakir   University of Burgundy, France
Smain Femmam   Université de Haute-Alsace, France
Zhiliang Qin   Shandong University, China
Wasana Boonsong   Rajamangala University of Technology Srivijaya, Thailand

Christoph Lange	HTW Berlin (University of Applied Sciences for Engineering and Economics), Germany
Tai-Myoung Chung	Sungkyunkwan University, South Korea
Noor Zaman Jhanjhi	Taylor's University, Malaysia
Mandeep Singh Jit Singh	Universiti Kebangsaan Malaysia, Malaysia
Natthanan Promsuk	Chiang Mai University, Thailand
Vojtech Uher	VSB-Technical University of Ostrava, Czech Republic
Rachid Oucheikh	Jönköping University, Sweden
Abidalrahman Mohd	Eastern Illinois University, USA
Jun Wu	Hangzhou Dianzi University, China
Agris Nikitenko	Riga Technical University, Latvia
Haya Abdullah Alhakbani	Al-Imam Muhammad Ibn Saud Islamic University, Saudi Arabia
Sara Blanc Clavero	Universidad Politécnica de Valencia, Spain
Zhifeng Wang	Signifyd, USA
Paolo Garza	Polytechnic University of Turin, Italy
Calin Ciufudean	Stefan cel Mare University of Suceava, Romania
Cuong Pham-Quoc	Ho Chi Minh City University of Technology, Vietnam
S. Akhila	B. M. S. College of Engineering (BMSCE), India
Vu Khanh Quy	Hung Yen University of Technology and Education, Vietnam
Ngoc Thanh Nguyen	Wroclaw University of Science and Technology, Poland
Viorel Nicolau	"Dunarea de Jos" University of Galati, Romania
George Petrea	"Dunarea de Jos" University of Galati, Romania
Pius Adewale Owolawi	Tshwane University of Technology, South Africa

Co-sponsors

 Universitatea
Ştefan cel Mare
Suceava

 Institut Teknologi
Telkom
Purwokerto

 UNIVERSITY
OF PARDUBICE
FACULTY
OF ECONOMICS
AND ADMINISTRATION

Contents

Computer and Electronic Engineering

Information Science and Mobile Communication

Towards Intelligent Management of Internet of Modern Drones

Supadchaya Puangpontip$^{(\boxtimes)}$ [ID] and Rattikorn Hewett[ID]

Texas Tech University, Lubbock, TX 79409, USA
{supadchaya.puangpontip,rattikorn.hewett}@ttu.edu

Abstract. Internet of drones (IoD) provides coordinated access, between drones and users, over the Internet to controlled airspace. With advanced drone, mobile and Artificial Intelligence (AI) technologies, today's drones are equipped with sophisticated onboard AI software that enhances drone services and our way of life (e.g., package delivery, traffic surveillance). As IoD grows, there is a need to effectively manage large-scaled drones with multiple regulation and resource constraints, particularly energy usage. This paper presents preliminary work on generic architecture and operations to lay foundations for intelligent drone management systems. By also introducing a method to pre-determine estimated energy consumption of deep neural net image analysis deployed in drones, the paper illustrates this work on managing the search rescue drone autonomy to decide on its actions based on energy consumption. The proposed approach can be extended to manage a network of drones and additional resource constraints including response time, safety or environmental compliance and financial budget.

Keywords: IoD · Energy consumption · Deep learning · CNN · Drone

1 Introduction

Internet of Things (IoT) is a key enabler to services that improve quality of our life. Evolving from IoT, Internet of drones (IoD) [1, 2] provides an infrastructure for coordinated access, between drones and users, over the Internet to controlled airspace. With advanced mobile and AI (Artificial Intelligence) technologies, modern drones are more affordable and equipped with sophisticated onboard smart software to enhance drone capabilities (e.g., image recognition with deep neural net learning (DNN), navigation). The applications of IoD are ample from traffic surveillance to data collection, to package delivery to disaster mitigation and rescue [1, 3]. There is an increasing need to effectively manage large-scaled drones with multiple resource (e.g., time, airspace, cost) and regulation (e.g., environmental, safety) constraints, particularly energy usage. Without good planning and energy management, the drone can exhaust its battery before it gets to the destination or finish the mission.

Recent research in IoD deals with energy-efficient solutions e.g., optimizing transmission power or flight time [4–6], harvesting energy or managing charging stations [3] to increase the drone's operating time. Little work has been done on energy management of the drones or energy modeling of the software, such as DNN, deployed in drones [3, 4, 7]. This paper presents preliminary work on generic architecture and operations for intelligent drone management systems. The "intelligence" is contributed by its adaptiveness to dynamically changing conditions and constraints. The paper also proposes a method to estimate energy consumption of DNN image analysis deployed in drones and illustrates its use in managing the rescue drones to recommend appropriate actions based on the remained energy. Unlike previous energy modeling [7], ours gives a method for deriving energy measurements from the DNN configurations.

The rest of the paper is organized as follows. Sections 2 and 3 are our main contributions of the proposed framework and drone energy models, respectively. The paper illustrates and experiments on rescue drones in Sect. 4 and concludes in Sect. 5.

2   The Proposed Framework for Intelligent Drone Management

Consider a system of modern drones of various sizes, functions (e.g., capture image or video, fly) and capabilities. For example, some are equipped with intelligent software on board (e.g., to detect certain objects from an image captured by the drone, navigate, or decide whether to fly back to the base or continue the excursion). Some may be able to fly at high speed but for a short duration, whereas some may have a long battery life but can only fly at medium to low speed. Our objective is to develop a framework for building an intelligent system (e.g., adaptive to changing and uncertain situations) to assist management of drones on a particular mission (e.g., drug delivery in rural area, surveillance, or rescue search). In this paper, we focus on managing the drone autonomy on its actions based on limited resource (i.e., energy). To convey the idea clearly, we pick a case scenario of using drones to help find victims from a disaster (e.g., flood or wildfires). Although we have not done this, we conjecture that the proposed approach can be extended to manage a team of drones and additional resource constraints (e.g., time, safety or environmental compliance and budget).

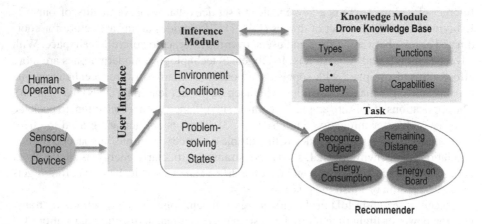

Fig. 1. An overall architecture of drone management framework.

We will refer to the drone management system employing the proposed architecture as RESCUER. Figure 1 shows an overview of RESCUER's proposed architecture that consists of three basic components: *Inference Module, Knowledge Module* and *Task Module* (e.g., *Recommender*). The architecture is data/event driven where the data are transmitted from external environments including sensors (e.g., temperature, weather/wind condition), human operators/users, and drone devices (e.g., image or video), as shown to the left of Fig. 1. The data interact with RESCUER via its User Interface to a reasoning module during problem solving. Each component is described below.

Inference Module: Includes an inference engine that provides basic reasoning mechanisms (e.g., forward or back-ward inferences) and a working module to maintain problemsolving states of reasoning tasks. In each reasoning cycle, recent each event/data triggers applicable reasoning operations from various tasks and select the most appropriate one for execution, which in turn would produce new events initiating a new reasoning cycle. Control mechanisms to select rules to act can use priorities, or meta-control rules [8].

Knowledge Module: Represents information about drones including type specifications and models, functions and requirements (e.g., transmission), capabilities, and battery life. These are used by the task module to assist drones to take appropriate actions. Note that some of the knowledge can be generic (e.g., physics of drone flying) and some are specific to task applications.

Task Module: Represents several computing tasks required or influence the recommendation to the drone. For example, by *"recognize object"*, the drone analyzes the image captured (using deep learning technology to recognize a target that it is searching). If the result is positive, then it will send the image to RESCUER to send a team to rescue the victim. This helps the drone to save transmission energy from transmitting all images captured. The "energy consumption" quantifies energy usage so far, while the "energy on board" calculates remaining energy that the drone has. Together with the "remaining distance" to travel, the recommender can recommend the drone to 1) fly back (if not enough energy, or environmental condition is not safe), or 2) continue the search (if enough energy and more area to search, 3) transmit the image that detects a target victim, and 4) alarm and increase speed when in danger.

Note that the proposed framework is general in that although the framework is applied to drone rescuers, it is also applicable to other IoT application domains.

3 Energy Estimation of Drone's Activities

This section describes a method to estimate energy consumption of different activities of the drone. Drone actions (e.g., analyze image, transmit image, go back, continue search, alarm) are driven by its awareness of remaining energy and the mission. The total consumption of the drone is the summation of the consumptions of all activities. Table 1 summarizes all but one of basic drone functions have estimated energy usage based on published work [4, 9].

Table 1. Energy consumption of some basic functions of drones.

Drone functions	Energy consumption	Variable descriptions
Hovering [4]	$((mg)^3/2\pi r_p^2 n_p \rho)^{1/2} t$	m: mass of the drone, g: gravitational acceleration, r_p: radius of the propellers, n_p: number of the propellers, and ρ: air density
Transition [4] (Flying)	$\sum_{i=1}^{n-1} (p_{full}/v_{full}) d_{i,i+1}$	p_{full}: hardware power of the drone at full speed, v_{full}: full speed, n: #locations, $d_{i,i+1}$ distance from location i to $i+1$
Transmission [10]	$(\eta p + p_s)(s/(b \log(1 + (pG/N_0 b)))$	η: coeff. of transmission power, p: transmission power, p_s power of the drone when no transmission, s: data size to transmit, b: bandwidth, G: channel power gain between the drone and the gateway, N_0: power spectral density

3.1 Hovering Energy

The hovering energy (E_{hov}) is the energy the drone consumes while remaining stationary in the air [11]. It depends on hovering power p_{hov} and transmission time t, shown in (1).

$$E_{hov} = p_{hov} t \tag{1}$$

$$p_{hov} = ((mg)^3/2\pi r_p^2 n_p \rho)^{1/2} \tag{2}$$

The hovering power can be obtained by using (2) where m is the mass of the drone, g is the gravitational acceleration, r_p is the radius of the propellers, n_p is number of the propellers, and ρ is the air density [4, 11].

3.2 Transition Energy

The transition (or flying; E_{fly}) energy is the energy the drone consumes during moving from one location to another. This can be obtained from hardware power of the drone p_{full} when moving at the full speed v_{full} and distance $d_{i,i+1}$ between the location i and location $i + 1$ [4]. Given n locations, the transition energy can be found using (3).

$$E_{fly} = \sum_{i=1}^{n-1} (p_{full}/v_{full}) d_{i,i+1} \tag{3}$$

3.3 Transmission Energy

Transmission energy (E_{trans}) of the drone depends on transmission power p and transmission time t [10]. The latter can be obtained by dividing the total number of bits to be transmitted s by the transmission rate r. This gives,

$$E_{trans} = pt = p(s/r) \qquad (4)$$

The transmission rate r can be obtained from (5) where b is bandwidth, p is transmission power, h is channel power gain between the drone and the IoD gateway and N_0 is power spectral density of the Gaussian noise [6, 9, 10, 12].

$$r = b \, \log(1 + (pG/N_0 b)) \qquad (5)$$

3.4 Software Energy

While estimating software energy can be easily done via electrical power measurement, when designing a system with limited resource, the ability to pre-estimate energy consumption without having to run them can be very useful.

Energy consumption of software is from computation (E_{comp}) and data movement (E_{data}). E_{comp} is αn_{MACs}, for a constant α and n_{MACs} number of MAC (*multiply-and-accumulate*) operations [7, 13]. For example, to compute $p = \sum_{i=0}^{n} w_i x_i$, each iteration i takes one MAC operation. With n iterations, it takes n MACs. On the other hand, one MAC requires three data reads (for w_i, x_i and p_{i-1}) and one write (i.e., new partial result p_i). Suppose data moves between two memory levels: *cache* and *DRAM* with a cache hit rate h. Data are first looked up in the cache and if they are not found (cache miss), they will be fetched from DRAM and store in cache. As a result, we can obtain data movement energy to be: $E_{data} = \Sigma_{v \in V} (\beta_{cache} a_v + \beta_{DRAM}(1 - h)a_v)p$ where β_m is a *hardware energy cost per data access* in m memory and V is a set of data (e.g., input, output, weight), a_v is the

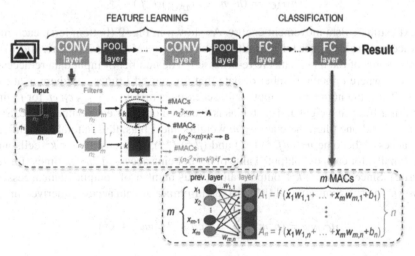

Fig. 2. Computation and associated data in CONV layer (left) and FC layer (right).

number of data accesses for data of type v, and p, precision of data in bits (e.g., 8, 16). Thus, total energy $E_{software}$ is shown in (6). More details are in [14].

$$E_{software} = \alpha\, n_{MACs} + \Sigma_{v \in V}(\beta_{cache}a_v + \beta_{DRAM}(1-h)a_v)p \qquad (6)$$

We propose a method to estimate energy consumption for image analysis that applies a popular trained DNN (Deep Neural Net) model, specifically the Convolutional Neural Net (CNN) architecture, to detect a human (victim) object in the captured image. The proposed energy model is different from our previous model [14]. However, both are based on the same logic. As shown at the top of Fig. 2, the flow of computation in CNN passes through multiple *convolution* (CONV) layers to extract features, *Pooling* (POOL) to reduce the dimensions and *fully connected* (FC) layers for classification. The basic computation of CONV and FC layers are shown at the bottom left and right of Fig. 2, respectively. Due to space limitation, we will omit POOL layer and the derivation of the number of MACs and data accesses for modeling energy in the CONV layer whose details can be found in [4]. Below are energy models with dimensions of input, filter and output which are specified in Fig. 2 (e.g., n_1, n_2, k, f, etc. in CONV).

CONV Layer Energy. First, we find n_{MACs}. As shown in Fig. 2 (bottom left), each CONV layer takes data input of size $n_1 \times n_1 \times m$ and convolves them with each of the f filters of size $n_2 \times n_2 \times m$ to give a final output of size $k \times k \times f$. The convolution process starts with a pointwise multiplication of an input plane with the filter plane of the same dimension (i.e., $n_2 \times n_2 \times m$). This step takes n_2^2 MACs. Then the results of each plane are summed for each of m channels. Thus, a total number of MACs required for convolving just one cell in this step is $n_2^2 m$ (as denoted by A in Fig. 2). Since there are k^2 cells for each filter and f filters, and thus it takes a total of $(n_2^2 m)(k^2 f)$ MACs (as denoted by B and C as a final count in Fig. 2). However, since each convoluted cell is fed to an activation function whose computation requires n_{MACs}^a MACs. With this additional computation, the CONV layer requires a total of $(n_2^2 m + n_{MACs}^a)(k^2 f)$ MACs. Hence, we obtain:

$$n_{MACs} = (n_2^2 m + n_{MACs}^a)(k^2 f) \qquad (7)$$

Next estimate data movement energy. As shown in Fig. 2 (bottom left), each input data is accessed at least once for each filter requiring $n_1^2 m$ MACs. As the filter slides over the input, some of the input data are re-accessed. The number of input data re-accessed is $\sum_{i=2}^{t} c_i i$ where c_i is the number of data that are reused i times and there is at most t reuses. Thus, the number of "input" data accesses is $(n_1^2 m + \sum_{i=2}^{t} c_i i) \cdot f$ for f filters (**). Similarly, each weight and each bias is accessed once for each output value. Since for one cell and one filter, there are $n_2^2 m$ weights and 1 bias, thus, a total "weight" and "bias" accesses become $n_2^2 m(f k^2)$ (**) and $(f k^2)$ (**), respectively, for k^2 cells and f filters. Finally, for each of "output" value of $n_2^2 m$ being accessed $2 n_2^2 m$ times (for read and write). Since there are $k^2 f$ output values, thus number of "output" data accesses is $(2 n_2^2 m)(k^2 f)$ (**). Summing all (**)'s, a total number of data access is derived in (8).

$$a_{CONV} = \left(n_1^2 m + \sum_{i=2}^{t} c_i i + 3 n_2^2 m k^2 + k^2\right) f \qquad (8)$$

By substituting (7) and (8) in (6), a total energy consumption of a CONV layer can be obtained as shown below:

$$E_{CONV} = \alpha(n_2^2 m + n_{MACs}^a)k^2 f + p[\beta_{cache} a_{CONV} + (1 - h)\beta b_{DRAM} a_{CONV}] \qquad (9)$$

FC Layer Energy. Each FC layer takes m neurons from previous layer. As shown Fig. 2, for each of n neurons in the FC layer, we compute m weighted sums (argument of an activation function f). Thus, the total number of MACs in the FC layer is shown in (10) where n_{MACs}^a is the number of MACs used in the activation function (e.g., Sigmoid takes one MAC (for a division) with extra computation γ for exponential function).

$$n_{MACs} = n(m + n_{MACs}^a) \qquad (10)$$

For data movement energy, we consider m input neurons, mn weights, n biases and n output neurons in FC layer (or output layer). As shown in computation of A_i's in Fig. 2, each input x_i is read n times while each weight and each bias are each read once. Thus, a total number of data accesses for input, weight and bias would be mn, mn, and n, respectively. Each output A_i includes read/write accesses of m products, yielding 2 m data accesses. Since there are n output neurons, a total of number of data accesses for output would be 2 mn. As a result, Total number of data accesses is obtained below.

$$a_{FC} = 4\,mn + n \qquad (11)$$

By substituting (10) and (11) in (6), we obtain total energy of the FC layer as shown in (12).

$$E_{FC} = \alpha n(m + n_{MACs}^a) + p(\beta_{cache}(4\,mn + n) + \beta_{DRAM}(1 - h)(4\,mn + n)) \qquad (12)$$

4 Search and Rescue Drones: Experiments and Results

Consider a scenario of the IoD on a rescue mission, we set up simulations based on the RESCUER management system developed from the framework architecture described in Sect. 2. The IoD has 10 drones, each of which takes 10 images. Comparing two drones, one is a typical drone where all images captured will be transmitted whereas the other is a RESCUER (or smart) drone that analyzes the image and transmits on if victims are detected.

Table 2 shows that the typical drone consumes more transmission energy and has higher total energy consumption. Consequently, it covers less distance and less number of locations. The smart drone, on the other hand, performs better. The ability to select what images to send allows the drone to save energy and spends them on searching. In this particular instance, the transmission energy is reduced by about 70% and the drone can visit 65% more locations. Figure 3 shows average energy breakdowns of 10 smart drones that detect victims (thus, image transmission) randomly with probability of 0.3. On the average, energy consumption is highest by software (or DNN) and lowest by flying. Image analysis task takes more energy on the average than transmission.

Table 2. Typical vs. RESCUER drones.

	Typical	RESCUER
Total Energy (J)	1922.94	1129.97
Transmission Energy (J)	1173.10	349.4
Software (DNN) Energy (J)	0	484.36
Hovering Energy (J)	651.48	194.04
Flying Energy (J)	98.36	102.17

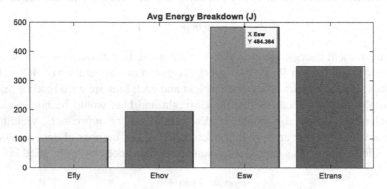

Fig. 3. Average energy breakdown by activities.

In addition, we experiment how varying hop distances will impact energy consumption. As shown in Fig. 4, for 100 m, the drones consume highest energy that gradually decreases for longer hop length. Since the drone is assumed to take pictures, analyze, and possibly transmit the pictures at every location, shorter hop distance means it has to perform these tasks more frequently and marks more locations. As a result, the total energy spent on these tasks is more than on traveling. Reversely, for 500 m, the drone covers more distance and consumes less energy. They however visit, on average, about

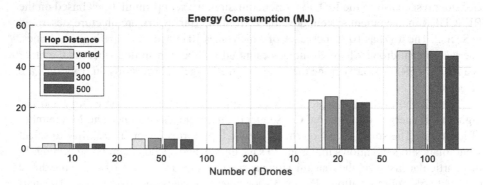

Fig. 4. Effects of varying hop distances.

280 less locations than that of 100 m setting. The randomly varied hop distance consumes about the same energy as those of the 300 m hop distance. This may be because average of the varied distance is close to 300 m.

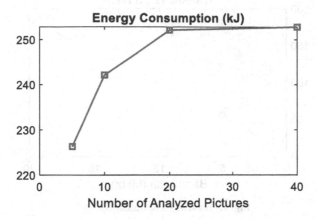

Fig. 5. Image analysis & energy

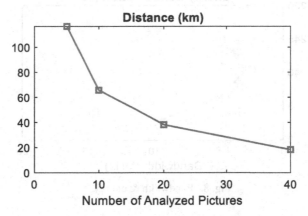

Fig. 6. Image analysis & distance

Figures 5–8 investigates relationships among relevant factors that may impact the mission including the number of images analyzed, distance, bandwidth, transmission rate and energy consumption. Number of analyzed pictures directly affect software energy and transmission energy, since it is the number of times the drone runs a DNN for victim detection and decides whether to transmit the image or not. On the other hand, bandwidth is a network resource which is not always plentiful. It can directly influence the drone's transmission and hovering energy.

As shown in Fig. 5, suppose the drone analyzes 5 images at every location, the total energy is about 230 kJ compared to analyzing 20 images with 250 kJ of energy. The more the images are analyzed, the more energy is consumed as expected. One may argue at the difference of 20 kJ may not be much, however, the distance covered can reduce significantly.

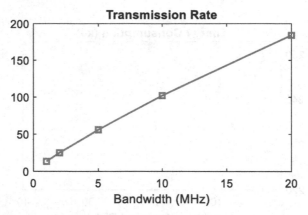

Fig. 7. Bandwidth & transmission.

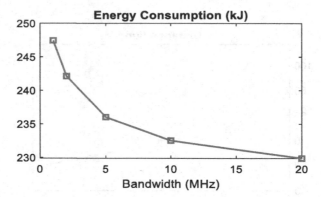

Fig. 8. Bandwidth & energy.

As shown in Fig. 6, the drone can travel about 113, 66, 35 and 18 km, with 5, 10, 20 and 40 analyzed images, respectively. The number of image processing impacts the distance traveled because energy is spent more on software (and transmission), less are spent on the travel. Figure 7 shows that high bandwidth gives high transmission rate. This in turn reduces time to transmit, thus, transmission energy, hovering energy, and the total energy consumption as shown in Fig. 8.

5 Conclusion

This paper presents a generic framework for building intelligent drone management systems. We also present a method for estimating energy consumption of drone functions, particularly the DNN deployed for object recognition in IoD and uses it in our illustration on RESCUER, an intelligent drone management. By focusing on the drone autonomy, RESCUER recommends appropriate actions to drones based on the remaining energy. Although we use simple scenarios and the experiments are preliminary, the results conform with the variable effects as anticipated. Our future research includes extending the framework to manage a network of drones, incorporating security and privacy aspects, and accounting for additional resource constraints e.g., response time, safety or environmental compliance and financial budget.

References

1. Abualigah, L., Diabat, A., Sumari, P., Gandomi, A.H.: Applications, deployments, and integration of Internet of Drones (IoD): a review. IEEE Sens. J. **21**, 25532–25546 (2021). https://doi.org/10.1109/JSEN.2021.3114266
2. Boccadoro, P., Striccoli, D., Grieco, L.A.: An extensive survey on the Internet of Drones. Ad Hoc Netw. **122**, 102600 (2021). https://doi.org/10.1016/j.adhoc.2021.102600
3. Alsamhi, S.H., Ma, O., Ansari, M.S., Almalki, F.A.: Survey on collaborative smart drones and internet of things for improving smartness of smart cities. IEEE Access **7**, 128125–128152 (2019). https://doi.org/10.1109/ACCESS.2019.2934998
4. Yao, J., Ansari, N.: QoS-aware power control in internet of drones for data collection service. IEEE Trans. Veh. Technol. **68**, 6649–6656 (2019). https://doi.org/10.1109/TVT.2019.2915270
5. Sarkar, S., Khare, S., Totaro, M.W., Kumar, A.: A novel energy aware secure internet of drones design: ESIoD. In: IEEE INFOCOM 2021 - IEEE Conference on Computer Communications Workshops (INFOCOM WKSHPS), pp. 1–6 (2021). https://doi.org/10.1109/INFOCOMWKSHPS51825.2021.9484461
6. Wang, L., Hu, B., Chen, S.: Energy efficient placement of a drone base station for minimum required transmit power. IEEE Wirel. Commun. Lett. **9**, 2010–2014 (2020). https://doi.org/10.1109/LWC.2018.2808957
7. Yang, T.J., Chen, Y.H., Emer, J., Sze, V.: A method to estimate the energy consumption of deep neural networks. In: Conference Record of 51st Asilomar Conference on Signals, Systems and Computers, ACSSC 2017, 2017 October, pp. 1916–1920 (2018). https://doi.org/10.1109/ACSSC.2017.8335698
8. Russell, S., Norvig, P.: AI a modern approach. Learning **2**, 4 (2005)
9. Mozaffari, M., Saad, W., Bennis, M., Debbah, M.: Drone small cells in the clouds: design, deployment and performance analysis. In: 2015 IEEE Global Communications Conference (GLOBECOM), pp. 1–6 (2015). https://doi.org/10.1109/GLOCOM.2015.7417609
10. Auer, G., et al.: How much energy is needed to run a wireless network? IEEE Wirel. Commun. **18**, 40–49 (2011). https://doi.org/10.1109/MWC.2011.6056691
11. Gundlach, J., Gundlach, J.: Designing unmanned aircraft systems: a comprehensive approach. American Institute of Aeronautics and Astronautics Reston, VA (2012)
12. Duangsuwan, S., Maw, M.M.: Comparison of path loss prediction models for UAV and IoT air-to-ground communication system in rural precision farming environment. J. Commun. **16**, 60–66 (2021)
13. Goodfellow, I., Bengio, Y., Courville, A.: Deep Learning. MIT Press, Cambridge (2016)
14. Puangpontip, S., Hewett, R.: Energy usage of deep learning in smart cities. In: Proceedings - 2020 International Conference on Computational Science and Computational Intelligence, CSCI 2020, pp. 1143–1148 (2020). https://doi.org/10.1109/CSCI51800.2020.00214

Ordering Algorithm as a Support for Children with ADHD Through the Development of Bilingual and Interactive Videogames

Roberto-Alejandro Barahona-Cevallos$^{(\boxtimes)}$ ⓘ, Johnny-Andrés Moya-Suárezⓘ, Milton-Patricio Navas-Moyaⓘ, and Ximena-Rocio López-Chicoⓘ

Universidad de las Fuerzas Armadas ESPE, Sangolqui, Ecuador
{rabarahona1,jamoya4,mpnavas,xrlopez}@espe.edu.ec

Abstract. Attention Deficit Hyperactivity Disorder (ADHD) involves a pattern of attention deficit, hyperactivity, and impulsivity. There are several works, where videogames are used as a technological support tool in the classroom. However, although the benefits are documented, it is not clear how it was created and especially the purpose of learning for which they were created, so the objective is: to make an application of the software named JaraSmart, in the use of ordering algorithm to support children with ADHD through the development of bilingual videogames. The results are encouraging and allowed us to establish a more formal basis for the use of bilingual videogames with the application of sorting algorithms as support for children with ADHD, aged 10 to 12 years .

Keywords: ADHD · Bilingual videogames · Hyperactivity · Motor skills · Cognitive

1 Introduction

Attention Deficit Hyperactivity Disorder (ADHD) is a pathology whose origin and the factors that produce it are unknown. Due to this, the Ministry of Education along with the Specialized and Inclusive Directorate have found 7918 students in the public system of Ecuador with this pathology. [1] Therefore, ADHD has a high psycho-social impact, which is reflected in the deterioration of the child's functioning in family, school, and social life.

For this reason, videogames have great advantages in the educational field, among which stand out the improvement in vision, increase in self-esteem, it favors an interactive learning, while promoting learning through challenge, it also allows improvement in social skills, language, reading rules and messages, and basic mathematics, as well as the articulation of an abstract thinking [2, 3].

Videogames with modern technology have a great applicability in psychology, and they greatly improve the daily lives of many people. New technologies are a perfect ally to treat difficulties, through different applications they can help improve attention, memory, and hyperactivity in children with ADHD in a much more enjoyable and fun way [4].

F. Neri et al. (Eds.): CCCE 2022, CCIS 1630, pp. 14–25, 2022.
https://doi.org/10.1007/978-3-031-17422-3_2

There is an application called EndeavorRx which is a video game that was approved by the FDA and aimed at children between 8 and 12 years old with ADHD, where it can be used along with other treatment options, including educational programs, medication and physician-directed therapy [5]. Through this example we can have a starting point for the application that has been created, where 6 different games have been developed that focus on attention and memory. With JaraSmart, the children, besides playing and developing their memory, are also learning mathematics, spelling and Ecuadorian culture. With this, children can have several benefits and at the same time the results obtained can be evaluated, so that experts can use it as an alternative treatment for ADHD.

To contribute to the latter, it focuses on the design-implementation-use-application system, which is called JaraSmart, and its main objective is to help children between 10 and 12 years old with ADHD, to improve memory difficulties, attention or hyperactivity through simple mini videogames adapted to their ages.

2   State of Art

Considering the factors of the advancement of technology and its insertion in everyday life, it is important to think about designing an application that helps 12-year-old children with ADHD, giving them a tool to improve their attention, using ordering algorithm to support them, through the development of bilingual videogames, with a set of tools that will allow the development of applications and software systems. Making it, in this way, an innovative and entertaining application especially for children [6].

2.1   C#

It is a general-purpose language designed by Microsoft for its platform. [7] Along with Java, it is one of the most used programming languages, used for the development of applications, providing the programmer with a level of abstraction that will be useful in the development of complex applications.

2.2   MeISE Methodology

For the development of the application, the methodology of Educational Software Engineering (MeISE) was used. [8] which exposes a life cycle divided into two stages: definition and development, as shown in Fig. 1.

The MeISE methodology bases its operation on the definition of the requirements based on the conceptual phase where the initial design of the application is sketched, the referential framework of the design is established, making sure that the product that is released in each plan is didactically complete [8].

In the second step, the software is developed with the technical team and the interaction with the specialists, making each interaction and designing the application, considering that they have a feedback process, to discover and proactively improve the difficulties presented in the teaching-learning process [9, 10].

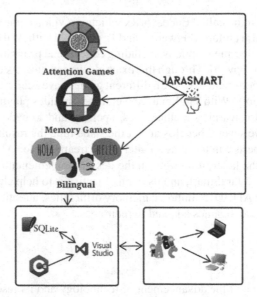

Fig. 1. Outline of the MeISE methodology.

3 Implementation

The educational software has a great impact on children with pathologies such as ADHD, so we used the requirements in order of priority depending on the effort required to develop it, this, through the MeISE methodology, that is in the field of Software Engineering [11], where it has a life cycle for development, consisting of 2 stages of definition that will help the proper implementation of the activities in a structured way.

The interactive videogame is titled "JaraSmart", its purpose is learning through challenge, promoting competition, improving social skills in language in a bilingual way, reading rules and messages, as well as the practice of basic mathematics and the articulation of abstract thinking as stipulated by educational quality [12, 13] (Fig. 2).

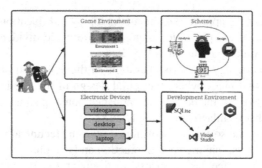

Fig. 2. Implementation schematic.

Based on the proposed context, using the MeISE method, different schemes are designed that provide an interactive approach and include informatics, pedagogical and communication aspects. This methodology proposes a cycle divided into two phases, the main one consisting of a specification and a preliminary drawing addressing the identified educational conditions.

The MeISE methodology for video games makes it possible to obtain a high-quality product, from a technical and pedagogical point of view, which includes instructional design and human-machine interface aspects. Therefore, auditory and visual stimuli are presented when there are encouraging messages such as "You advance in level", "Joker", "Play again", so that the child can see the success or failure, and be encouraged and try again, which is expected to promote healthy competition.

Once the main activities have been identified, the second phase begins, in which the development of different attention and memory video games is proposed, which are the most common areas affected by ADHD [14]. These focus on activities that stimulate the mind and concentration on letters, images, colors, as well as the development of mathematical, cultural and literary thinking according to the objectives established in the application.

The main advantage of MeISE is its focus on the design of video games and peripherals, as these devices contribute to the improvement of reading and writing skills in children with ADHD [15]. The technology provided by C# is used for the functionality of video games and interactive desktop applications. This language complemented with the IDE Visual Studio 2022, allow to define the functions of the video game from object-oriented programming modules and help to organize in folders, methods and functions, in addition to building the application. All the information of the scores of the different attention and memory videogames is stored in a local SQLite database that will be visible to each player, implementing the bubble sort algorithm to sort according to the best score [16] (Fig. 3).

3.1 Main Screens

Fig. 3. Startup screen.

On the home screen there is a field for the child to enter his or her name. This interface is made with the aim of being as simple as possible so that children are not distracted and can access in a quick and easy way. In addition, the language of the application can be changed from this screen (Fig. 4).

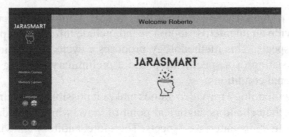

Fig. 4. Main screen.

The main screen consists of two parts, a left sidebar that contains the options menu and a panel that occupies the remaining space where the different screens and videogames will be loaded.

In the left menu you will be able to: enter the attention and memory videogames, change the language of the application, open the settings screen, and access the help screen.

3.2 Configuration Screen

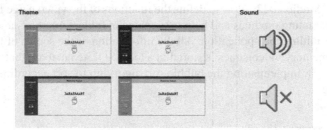

Fig. 5. Configuration screen.

In this interface you can change the main theme of the application, in which the child can choose the colors that he/she likes the most, so that he/she feels comfortable and entertained with the application. Here you can also enable or disable the sound depending on how it is needed (Fig. 5).

3.3 Help Screen

This application is designed to be as intuitive as possible, so there is a help screen where you can find different annotations so that the child or the professional in charge knows how to use this application (Fig. 6).

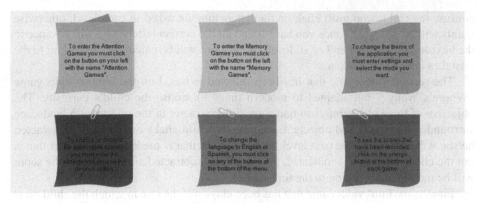

Fig. 6. Help screen.

3.4 Attention Games

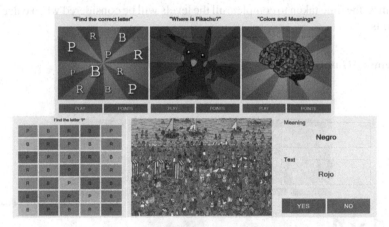

Fig. 7. Attention videogames screen.

In the attention screen (Fig. 7) you can see 3 different videogames that have been selected based on scientific studies and have been shown to develop the child's attention in different ways. Each videogame has different levels of increasing difficulty.

At the beginning of each videogame the objective and instructions will be shown in a clear and summarized way, after this, the time will start to run, to measure the performance of the child based on the successes and mistakes he has had. In addition, in each of the videogames you will have the option to give up if you are not able to finish it or a button that will help you with a hint depending on the videogame.

The first videogame is about the child having to choose the correct letter from a group of letters that are similar. The aim of this videogame is to develop the child's selective attention, which is the ability to focus on relevant information while ignoring irrelevant distractions. There will be shown several squares with a letter and similar colors to confuse the child, who must click on the letters that are asked in each level, otherwise points will be deducted. Once you have found all the correct letters, you will advance to the next level and at the end of all levels, your score will be calculated according to the mistakes and the time taken.

The second videogame that has been selected is based on the famous videogame "Where's Wally?" but adapted to modern times to arouse the child's curiosity. The objective of this videogame is to find a specific character in the middle of a landscape surrounded by people and objects. Each time the child clicks on the hidden character, he/she will advance to the next level. For each click that is made in another part that is not the character that was indicated, points will be subtracted, and at the end the score will be measured according to the time taken.

Finally, the third videogame that has been chosen is the one in which the child must decide if the color of the word below is the one that is written in the text above. Two words will be shown, one above and one below, if above is written 'Blue' and the word below is blue then the child must choose that it is correct. This videogame has a degree of difficulty in which it forces the child to be completely concentrated to achieve it. The fewer mistakes they make, the higher their score will be and as with the previous videogames, the time taken to complete all the levels will be considered when calculating their points.

3.5 Memory Games

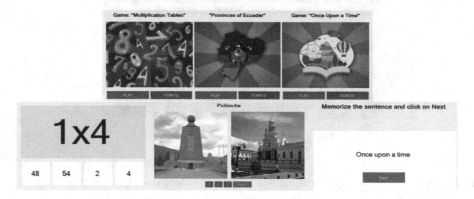

Fig. 8. Memory videogames screen.

In the memory screen (Fig. 8), there are 3 videogames that have been shown to help develop children's memory. Similarly, here, each videogame consists of different levels of increasing difficulty. In these videogames we have sought not only to develop the cognitive part of the child, but also to serve to learn and reinforce their knowledge, in a pleasant and healthy way.

The first videogame is geared towards mathematics, specifically the multiplication tables. The goal of this videogame is to improve quantitative reasoning and problem-solving skills that are necessary to make intelligent decisions, including those related to procedural memory. Here, multiplications will be randomly displayed and there will be 4 answers, 3 of which are random and 1 of which is correct. Therefore, the child will have to look for the option that is right, to score points. In case he/she gets it wrong, points will be deducted, and he/she will not be able to advance to the next multiplication. Also, the time advances every second to calculate the child's total score at the end of the 3 levels.

The second videogame is based on the provinces of Ecuador. The main objective of this videogame is to develop the child's memory, working with their temporal memories, through representative images. In addition, it is intended to get the child to learn about Ecuadorian culture. The videogame works as follows... In each level the child will be shown several images of 3 different provinces of Ecuador for the child to memorize them in the shortest time possible. After the child is ready, he will be shown a random image of each province and he will have to write down which province each of them belongs to. If the child makes a mistake, points will be deducted, and he/she will not be able to advance to the next level until he/she gets all 3 images right.

The third videogame is one of the most complex and is based on the videogame "once upon a time", where the child must memorize different sentences to from a story. The aim of this videogame is to develop short-term memory, as well as spatial memory, by making the child remember the order of the words that will be shown in different sentences that form a story. Here a sentence will appear, and when the child has memorized it, he will have to write it down, and concatenate it with the sentences he has memorized previously. Each time it becomes more and more complicated until he completes the little story. It is worth mentioning that here the child will also practice his spelling, since he will have to write the words correctly to advance.

3.6   Score Screen

Fig. 9. Scoring screen.

As can be seen in Fig. 7 and Fig. 8 below each videogame, there is an orange button, which will open the scores screen, where a table will be displayed with the results of all the children, sorted in descending order, using the bubble algorithm. This table consists of the position in which the child has been put, his name and the total points he has obtained (Fig. 9).

As you can see, there are two records with the same name, which means that a child can play several times, so that he can see all his scores and develop his healthy instinct for competition to improve.

4 Evaluation of Results

For this study, a quantitative analysis was executed focusing on the treatment of multi-dimensional interventions in ADHD [17], which focuses its attention on healthy competition through cognitive development of attention and memory. In this vein, it can be argued that quantitative analysis can be considered the best for understanding the use of ordering algorithm to support children with ADHD through the development of bilingual videogames. This study will be beneficial as it will critically evaluate both the positive and negative aspects of videogame-based therapy and how it can improve the situation of children affected by ADHD.

This study will be based on the collection of scores depending on the videogame being played and whether it focuses on attention or memory to achieve the outlined objectives. Bilingual videogame-based support can be considered a new phenomenon and therefore the opinion of different researchers was required instead of conducting primary research. Through an average maximum *score* of 1900 points in the attention videogames, where the following intervals will be considered (0–475 = difficult, 476–950 = average, 951–1425 = easy, 1426–1900 = domain) and 1100 in the memory videogames, whereby the following intervals will be considered (0–275 = difficult, 276–550 = average, 551–825 = easy, 826–1 100 = domain).

Table 1. Results of the average scores obtained in the different videogames.

Videogame	Attention					Memory				
Kid	Find the correct letter	Where is Pikachu?	Colors and meanings	Average score	Average % score	Multiplication tables	Provinces of Ecuador	Once upon a time	Average score	Average % score
1	1300	2600	1000	1633,30	85,96	900	800	700	800,00	72,73
2	1400	2800	1100	1766,70	92,98	800	900	800	833,30	75,75
3	1000	2200	900	1366,70	71,93	800	600	600	666,70	60,61
4	1300	2600	1000	1633,30	85,96	1100	900	800	933,30	84,85
5	1100	2100	900	1366,70	71,93	1000	800	800	866,70	78,79
6	1400	2200	1150	1583,30	83,33	1100	800	800	900,00	81,82
7	1300	1900	1100	1433,30	75,44	1100	700	700	833,30	75,75
8	1500	2700	1100	1766,70	92,98	1000	900	900	933,30	84,85
9	1400	2300	1000	1566,70	82,46	1100	800	800	900,00	81,82
10	1100	2500	900	1500,00	78,95	1000	900	600	833,30	75,75
11	1000	2100	800	1300,00	68,42	900	800	700	800,00	72,73
12	1000	2500	900	1466,70	77,19	1100	900	700	900,00	81,82
% - Total	**Max Average** = 1900				80,63	**Max Average** = 1100				77,27

Table 1 shows the results obtained by the children after having done 2 reviews and these were close to their nearest hundred for the ease of the study, showing an average of 80.63% success rate in the attention videogames and 77.27% success rate in the memory videogames.

Based on the results obtained, this application to support children with ADHD shows a great acceptance in the different types of videogames, as practice and dedication will make them obtain better results.

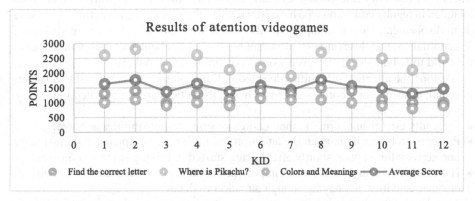

Fig. 10. Graph of the average scores of the attention videogames.

As can be seen in Fig. 10, all children have obtained quite favorable scores, although, child #3 had a lower score than the average. On the other hand, it can also be seen that the videogame "Where is Pikachu" was the most satisfactory, thus demonstrating that children with ADHD can focus their attention on a single thing when they set their minds to it.

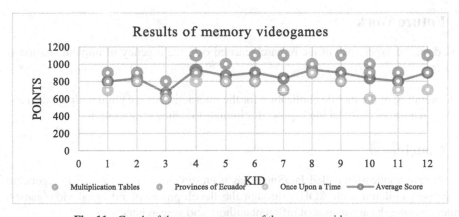

Fig. 11. Graph of the average scores of the memory videogames.

As can be seen in Fig. 11, all the children have obtained favorable scores; however, child 3 has had a little more difficulty with videogame 2, therefore, his percentage of success has been reduced. Furthermore, it can be deduced that the videogame in which the children have performed best is the multiplication tables, demonstrating that children with ADHD can solve mathematical exercises like the rest of the children.

5   Discussion

Children worldwide who suffer from ADHD require a lot of patience and control by teachers of educational institutions, in turn they do not know how to work with children included in regular education who have ADHD, which causes a negative attitude, making them disobedient, violent, compulsive, distracted, among others [18]. Being victims of discrimination, they lose interest in continuing to attend school activities, which causes the child to fail the grade, and even worse, to drop out of school [19].

Among the attention difficulties detected in Ecuador, they can manifest themselves in social, academic or laboral situations. These difficulties can be translated into:

- Difficulty establishing order in their chores or small responsibilities at home.
- Presents problems in maintaining attention until tasks are completed and must leave one activity for another, shortly after having started it, leaving several unfinished.
- Has trouble selecting what is most important, makes careless mistakes in schoolwork or other activities, not paying enough attention to details.
- They have difficulty paying attention to two alternative or simultaneous stimuli (e.g., listening to the teacher and taking notes at the same time).
- Often avoids or resists tasks that require sustained mental effort and/or a high degree of organization.

Its benefits are sought not only to aid memory and attention but also to have potentially long-lasting effects on other areas of the brain, the brain effects of cognitive development training have been shown to be sustained for five years [20, 21].

6   Future Work

Based on the experience of the research carried out, the urgency of implementation is clear:

- Generation of the mobile application for the interaction of the child with the devices.
- Application of artificial intelligence for human-machine interrelation

7   Conclusion

A software application called JaraSmart has been developed using ordering algorithm to support children with ADHD through the development of bilingual videogames, understanding the arguments of different authors and researchers.

Game-based training was provided to improve memory, attention, and impulsivity. Provided 6 progressive mini games with different difficulty levels and scores.

A results manager is provided that allows families and teachers to observe which exercise obtains better results for control and follow-up.

References

1. Villagómez Puebla, A.M.: Diagnosis and management of children with ADHD in Ecuador (2018)
2. Guerrero, J., González, J.: Videogames in special education: children with ADHD. Revista Digital AIPO-Asociación Interacción Persona-Ordenador. 2(1), 48–59 (2021)
3. Global Games Market Report: Newzoo's global games market report 2020. https://str ivesponsorship.coni/wp-content/uploads/2020/07/Global-Games-Market-Report-2020.pdf. Accessed 29 Nov 2021
4. Mayta, L., Rodríguez, G., Carvajal, B.: Videogame to improve attention in 10–12-year-old children with ADHD applying augmented reality concepts. Licentiate thesis, Universidad Mayor de San Andrés, Facultad de Ciencias Puras y Naturales Carrera de Informática, La Paz
5. Palazio-Arko, G.: Proceedings of the IX International Congress on Open Education and Technology (2016). https://addi.ehu.es/bitstream/handle/10810/25910/UCPDF164894.pdf? sequence=l&isAllowed=v. Accessed 29 Nov 2021
6. EcuRed: Using computer-assisted instruction (CAE) (2021). https://www.ecured.cu/Tut ores_inteligentes. Accessed 28 Nov 2021
7. Lucas Hoyos, S.C.: Educational software in C# to dynamize the active participation in language and literature class for the eighth grade (2021)
8. FIGUEROA MMAA: MeISE: Methodology of educational software engineering, 1116
9. Barragán, G.S.S., et al.: Mobile application for comprehension of mathematics in fourth grade children. Multidisciplines of Engineering
10. López, M., Rosero, E.: Description of the application of curricular adaptations in students with special educational needs not associated with a disability in the Unidad Educativa las Americas. Licentiate Thesis. Ambato: Technical University of Ambato, Faculty of Human Sciences and Education
11. Cabero, I., Barroso, J.: New digital scenarios. In: Information and Communication Technologies Applied to Training and Curriculum Development, 2nd edn. Editorial Piramide (2016)
12. UNESCO: Ministry of Education of Ecuador (2018). https://educacion.gob.ec/wp-content/ uploads/downloads/2013/07/Modulo_Trabajo_EI.pdf. Accessed 28 Nov 2021
13. Macias, K., Villafuerte, J.: Teaching English language in Ecuador: a review from the inclusive educational approach. J. Arts Hum. 12(2), 75–90 (2020)
14. Santurde del Arco, E.: Media education to improve academic performance and acquisition of digital competence in students with ADHD. University of Deusto, Madrid
15. Van Mechelen, M., Derboven, J., Laenen, A., Willems, B., Geerts, D., Abeele, V.V.: The GLID method: moving from design features to underlying values in co-design. Int. J. Hum. Comput. Stud. 97(1), 116–128 (2017)
16. Pariser, E.: The bubble filter: how the web decides what we read and what we think. Taurus (2017)
17. Miranda Casas, A., Soriano Ferrer, M.: Effective psychosocial treatments for attention deficit hyperactivity disorder. Psychol. Inf. 4(2), 110–114 (2016)
18. Rief, S.F.: How to Reach and Teach Children with ADD/ADHD: Practical Techniques, Strategies, and Interventions. 3rd edn. New York (2016)
19. Palacios-Cruz, L., de la Peña, F., Valderrama, A., Patiño, R., Calle Portugal, S.P., Ulloa, R.E.: Knowledge, beliefs, and attitudes in Mexican parents about attention deficit hyperactivity disorder (ADHD). Salud Mental. 34(2), 149–155 (2011)
20. Arroyo, S.: Brain training helps patients with ADHD. https://www.salud180.com/maternidad-e-infancia/entrenamiento-cerebral-favorece-los-pacientes-con-tdah
21. Simons, D.J., et al.: Do "brain-training" programs work? Psychol. Sci. Public Interest 17(3), 103–186 (2016)

Detection of Lateral Road Obstacles Based on the Haar Cascade Classification Method in Video Surveillance

Papa Assane Diop(✉), Amadou Dahirou Gueye, and Abdou Khadre Diop

TIC4DEV TEAM, Alioune DIOP University of Bambey, Bambey, Senegal
papaassane.diop1@uadb.edu.sn

Abstract. In this paper, we address the problem of improving driving safety on main roads outside built-up areas in sub-Saharan countries, particularly in Senegal, where there are more than 600 deaths per year in traffic accidents. The driver's field of vision in the rearview mirror, which is limited by the presence of a blind spot, does not always allow him to detect a potential obstacle in time. In this work, we propose a solution for real-time detection of some lateral obstacles on roads outside builtup areas from images acquired by video surveillance. To manage the detection of several obstacles in real time, we used the Haar cascade model to propose a sub-model for each type of obstacle. Thus, these models will be grouped in a single model. To achieve this we first collected data (images). The images collected must be of two types: positive images and negative images. The positive images must contain the obstacles. Negative images are images that do not contain obstacles. The evaluation of our model was highlighted through video image captures made on the road of Kaolack (a region of Senegal) with an accuracy rate of 88% determined from the ROC curve.

Keywords: Obstacle · Detection · Images · Video surveillance · Collection · Haar cascade · Lateral obstacles

1 Introduction

In developing countries, we are witnessing year after year a dramatic increase in the number of vehicles in regional capitals. In Senegal, the figures for the last 10 years reveal the following statistics: 5,024 deaths, since 2008 [1]. As the number of cars on the road increases, the number of accidents increases to more than 600 deaths each year [2]. With all these problems related to roads, video surveillance is a necessary means to ensure road safety. The video surveillance, commonly called video protection, consists of cameras and everything useful to record and exploit the images to detect abnormal events. The main objective of digital image processing is the extraction of information and the improvement of its visual quality in order to make it more interpretable by a human analyst or an autonomous machine perception.

Examples of digital images are those acquired by digital cameras, sensors on board satellites or aircrafts, medical equipment, industrial quality control equipment [3, 4].

F. Neri et al. (Eds.): CCCE 2022, CCIS 1630, pp. 26–35, 2022.
https://doi.org/10.1007/978-3-031-17422-3_3

The use of video surveillance allows us to use computer vision, which includes many classification models of lateral road obstacles to ensure road safety. In the literature, there are studies oriented to obstacle detection [1, 2, 5]. The haar cascade models are features used in computer vision for object detection in digital images. Very simple and fast to compute, they were used in the first real-time face detector, the Viola and Jones method [6]. After some years of work, these models can be used for other detection objects. Haar cascade is very useful on video surveillance more precisely on image and video processing. In this paper, we show that haar cascade models can present a reliable method on the detection of lateral road obstacles. However, the trained cascade haar model can only detect one type of obstacle. The rest of the paper will be structured as follows: in Sect. 2, we present a review of the literature on lateral road obstacles. In Sect. 3, we propose a classification model that relies on haar cascade and detects several types of lateral obstacles at the same time. In Sect. 4, we present the results obtained.

2 Review of Literature

In this section, we present some related work with respect to lateral obstacles. In [7], the term obstacle is taken to mean a hazardous obstacle. It refers to any object lateral (to the roadway), arrangement or fixed structure, whether point or continuous, that is likely to aggravate the consequences of a vehicle accidentally leaving the roadway, especially by causing a blockage or by promoting a rollover of the vehicle [5]. Obstacles can be divided by type [8].

2.1 Struck Obstacles

These are obstacles that are likely to be hit:

- Trees, which, with 37% of fatalities in obstacle collisions, certainly represent a major issue [7, 8];

Fig. 1. Tree obstacle.

Figure 1 shows trees considered as road obstacles. There is a lot of work on trees as in [9], which deals with the development of an approach to avoid trees in the path of drones in order to obtain a collision-free path [10].

- Electric or telephone poles that constitute 10% [7].

Fig. 2. Electric pole.

Steel utility poles, which are designed and manufactured to have a load capacity equivalent to that of wood poles according to the criteria [11]. The poles are therefore extremely hard which becomes a great obstacle to avoid.

- Masonry (12%), such as walls, parapets, aqueduct heads, supports for engineering structures, etc. [7].

2.2 Increased Risks in Curves

This type of obstacle is a particular hazard because the curve (bend) is an accident-prone feature. Figure 3 describes a bend in a heavily treed area.

Fig. 3. Increased risk in curves.

3 Classification Model

3.1 Haar Cascade Model

Haar's cascading classifier is one of the first and best-known object detection methods. This method, commonly known as the Violas and Jones method created by Paul Violas and Michael Jones, is one of the very first methods capable of effectively and real-time detecting objects in an image [4].

Viola and Jones' method involves scanning an image using a detection window of initial size 24 px by 24 px (in the original algorithm) and determining if a face is present on it. When the image has been browsed entirely, the size of the window is increased and the scan starts again, until the window is the size of the image. The increase in the size of the window is done by a multiplicative factor of 1.25. Swiping, on the other hand, is simply shifting the window by one pixel. This offset can be changed in order to speed up the process, but a one-pixel offset ensures maximum accuracy. This method is an appearance-based approach, which involves going through the entire image by calculating a number of features in overlapping rectangular areas. It has the particularity of using very simple but very numerous characteristics. There are other methods but that of Viola and Jones is the most efficient at the moment. What differentiates it from the others is in particular: the use of integral images that make it possible to calculate more quickly the selection by boosting the characteristics. The cascading combination of boosted classifiers, bringing a clear saving in execution time.

Learning the Classifier. A preliminary and very important step is the learning of the classifier. It is a question of training the classifier in order to make him aware of what we want to detect, here obstacles. For this, he is put in two situations: the first where a huge amount of positive cases are presented to him and the second where, conversely, a huge amount of negative cases are presented to him. Concretely, an image bank containing obstacles is reviewed in order to train the classifier. Then, an image bank containing no obstacles passed. In this case, Viola and Jones trained their classifier using an image bank. The result is an obstacle-sensitive classifier. It comes in the form of an XML file. In absolute terms, we would be able to detect any distinctive sign from a classifier trained to do this.

Features. A feature is a synthetic and informative representation, calculated from pixel values. The characteristics used here are the pseudo-haar characteristics. They are calculated by the difference in the pixel sums of two or more adjacent rectangular areas. Let's take an example: Here are two adjacent rectangular areas, the first in white, the second in black. The characteristics would be calculated by subtracting the sum of the black pixels from the sum of the white pixels. Characteristics are calculated at all positions and scales in a small detection window, typically 24 × 24 pixels or 20 × 15 pixels. A

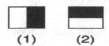

Fig. 4. Pseudo-haar characteristics.

very large number of features per window is thus generated, with Viola and Jones giving the example of a 24 × 24 size window that generates about 160,000 features.

The Fig. 4 has pseudo-haar characteristics with only two characteristics but there are others, ranging from 4 to 14, and with different orientations. Unfortunately, calculating these characteristics in the classic way is expensive in terms of CPU resources, which is where full images come in.

Classifier Cascade. Viola and Jones' method is based on a comprehensive image-wide research approach, which tests the presence of the object in a window at all positions and at multiple scales. However, this approach is extremely costly in calculation. One of the key ideas of the method to reduce this cost lies in the organization of the detection algorithm into a cascade of classifiers. Applied sequentially, these classifiers make a decision of acceptance, the window contains the object and the example is then passed to the next classifier, or reject, the window does not contain the object and in this case the example is permanently discarded. The idea is that since the vast majority of windows tested are negative (i.e. do not contain the object), it is advantageous to be able to reject them with as little calculation as possible. Here, the simplest, and therefore the fastest, classifiers are located at the beginning of the cascade, and very quickly reject the vast majority of negative examples. This cascading structure can also be interpreted as a degenerate decision tree, since each node has only ÷ one branch (Fig. 5).

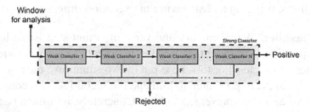

Fig. 5. Haar cascade architecture.

Illustration of the architecture of the waterfall: the windows are treated sequentially by the classifiers, and rejected immediately if the answer is negative One of the limitations of the method of Viola and Jones is its lack of robustness to rotation, and its difficulty in learning several views of the same object. In particular, it is difficult to obtain a classifier that can detect both faces from the front. Viola and Jones proposed an improvement that corrects this defect, which consists of learning a cascade dedicated to each orientation or view, and using a decision tree during detection to select the right waterfall to apply. Several other improvements were later proposed to address this issue.

Adaboost. Boosting selection involves using multiple cascaded "weak" classifiers rather than using a single "strong" classifier. Indeed, with a single classifier called "strong" that would present itself in this way (Fig. 6).

- Low classifier: performs at least better than random:

$$h(x, f, p, \theta) = 1, pf(x) > p\theta\,0, \text{otherwhise} \tag{1}$$

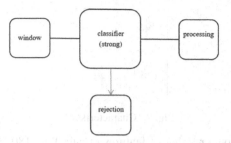

Fig. 6. Adaboost.

- Combine weak classifiers into a weighted sum to form a strong classifier:

$$C(x) = 1 \text{ si } \sum\nolimits_{(t=1)}^{T} \alpha_t h_t(x) \geq 1/2 \sum\nolimits_{(t=1)}^{T} a_t 0 \qquad (2)$$

3.2 Proposed Classification Model

The proposed model has several sub-models because for each road obstacle we need a model in its own right. Therefore, to use the haar cascade, there is an important step which is the collection of data (images). The images thus collected must be of two types: positive images and negative images. Positive images should contain obstacles. Negative images are images that do not contain obstacles (Fig. 7).

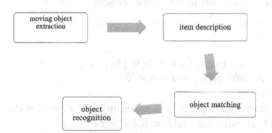

Fig. 7. Block diagram for object recognition.

The diagram above shows us the different for the creation of our detection method.

3.3 Extraction of Obstacle Characteristics

As soon as the image base is formed, we can extract the characteristics defined by Viola and Jones in their articles [6]. Haar's waterfalls create the features he uses by applying a 24 × 24 square filter to the image. These are the filters shown in the image below. Characteristics are obtained by subtracting the sum of the pixels in the image covered by the white area of the filter from the sum of the pixels covered by the blue area. This process generates a lot of characteristics since the filter is applied throughout the image.

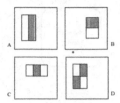

Fig. 8. Characteristic.

Viola and Jones mention a number of features greater than 180,000 in their article for an image of 384 by 288 [6] (Fig. 8).

Examples of rectangular entities shown against closed detection windows. The sum of the pixels that are in the white rectangles are subtracted from the sum of pixels in the gray rectangles. The characteristics with two rectangles are illustrated in (A) and (B). Figure (C) shows a rectangle with three characteristics, and (D) a characteristic with four rectangles [4, 6].

3.4 Classifier Training

For performance reasons, Viola and Jones chose to use the AdaBoost (Adaptive Boosting) classifier for the Haar detector [6]. Compared to other classifiers, this classifier has the advantage of having a good execution speed for the application that is made of it while maintaining a high recognition rate.

Algorithm 1: AdaBoost in the case of concept learning

Begin

$S = \{(x_1, u_1), \ldots, (x_m, u_m)\}$, with $u_i \in \{+1, -1\}$, $i = 1, m$
for all i=1,m make $p_0(x_i) \leftarrow 1/m$ $t \leftarrow 0$
for $t \leq T$ do
  Draw a learning sample S_t in S according to the probabilities p_t
  Learn a h_t classification rule on S_t by algorithm A
  Let ε_t be the apparent error of h_t on S. Calculate $\alpha_t \leftarrow 1/2 \ln (1 - \varepsilon_t/\varepsilon_t)$
  for all i = 1, m do
    $p_{t+1}(x_i) \leftarrow (p_t(x_i) /Z_t)e^{-\alpha t}$ if $h_t(x_i) = u_i$ (well classified by h_t)
    $p_{t+1}(x_i) \leftarrow (p_t(x_i) /Z_t)e^{+\alpha t}$ si $h_t(x_i) \neq u_i$ (misclassified by ht).
      (Z_t is a normalization value such that $\sum_{i=1}^{m} p_t(x_i) = 1$)
  end
  $t \leftarrow t + 1$
end For
Provide the final hypothesis: $H(x) = sign(\sum_{t=1}^{T} \alpha_t h_t(x)$

End

Adaboost [12].

The adaboost algorithm is implemented in many programming languages such as python in the Opencv model. One of the ideas is to define at each of its stages $1 \leq t \leq$

T, a new distribution Dt of a priori probabilities on the learning examples according to the results of the algorithm in the previous step. The weight in step t of an example (x_i, u_i) of index i is denoted $p_t(i)$. Initially, all examples have the same weight, and then at each stage, the weights of the examples misclassified by the learner are increased, thus forcing the learner to focus on the difficult examples in the learning sample.

4   Results Achieved

The videos were recorded from a camera on the road to Kaolack. As shown in Fig. 2, we captured images at each well-defined time in order to be able to evaluate the model (Fig. 9).

Fig. 9. Capturing obstacles to detect.

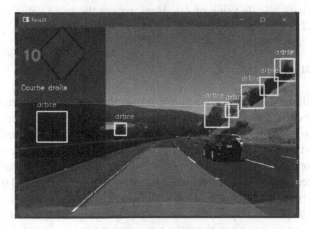

Fig. 10. Detection result.

We can see that the model detects trees as a side road obstacle. For the following, we will calculate some measures like recall and accuracy obtained from true positives and false positives, with a different scale factor value to determine its best value giving the most performance classification (model) (Fig. 11).

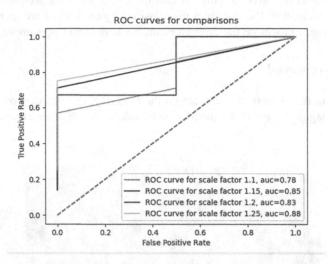

Fig. 11. ROC curves for comparisons.

The receptor functioning characteristics curve (ROC) is a probability curve that illustrates the quality of our binary classification consists of classifying subjects according to true positives and false positive rates. The area under the curve (AUC) is a metric that ranges from 0 to 1 [13]. AUC is a metric that evaluates the extent to which a logistic regression model ranks positive and negative outcomes as much as possible. In this paper, Fig. 10 shows the ROC for the classification of obstacles (trees). The AUC accuracy is 78% first curve with a scale value of 1.1. Here we see the scale value that presents a better AUC is 1.25.

5 Conclusion

Side road obstacles are increasingly a problem over the years with the increase in cars. First, we presented the different types of road side obstacles in order to propose a solution based on the haar cascade classification in video surveillance. With this method, we were able to detect obstacles by having the care to choose the scale factor. In the perspective of the work, we propose to go further to find the approximate distance of the obstacle and also to use another classification tool allowing a better detection of lateral road obstacles and many other types of moving obstacles such as wild animals and humans crossing the roads. We also plan to improve the work by classifying the types of obstacles and sending information to the driver about the type of obstacle over a certain distance.

References

1. Side obstacle treatment. http://dtrf.setra.fr/pdf/pj/Dtrf/0003/Dtrf-0003110/DT3110.pdf. Accessed 4 June 2020
2. Codatu. https://www.codatu.org/actualites/les-accidents-de-la-circulation-dans-la-commune-de-mbour-au-senegal-memoire-master-lome/. Accessed 12 Nov 2021
3. Ramadhani, M.I., Minarno, A.E., Cahyono, E.B.: Vehicle classification using haar cascade classifier method in traffic surveillance system. Kinetik Game Technol. Inf. Syst. Comput. Netw. Comput. Electron. Control 3(1), 57 (2017)
4. POLYPUBLIE. https://publications.polymtl.ca/2326/. Accessed 3 Oct 2021
5. Wei, Y., Li, Y., Hu, S.: On-road obstacle detection based on stereovision analysis. In: 2007 International Conference on Mechatronics and Automation, pp. 958–962. IEEE, Harbin (2007). https://doi.org/10.1109/ICMA.2007.4303676
6. Viola, P., Jones, M.: Rapid object detection using a boosted cascade of simple features. In: Proceedings of the 2001 IEEE Computer Society Conference on Computer Vision and Pattern Recognition, CVPR 2001, pp. I-511–I-518. IEEE, Kauai (2001). https://doi.org/10.1109/CVPR.2001.990517
7. Cerama. https://www.cerema.fr/fr/centre-ressources/boutique/savoirs-base-securiteroutiere. Accessed 21 Sept 2021
8. Unitheque. https://www.unitheque.com/accidents-contre-obstacles-milieu-urbain/dossiers/certu/Livre/48867. Accessed 22 Oct 2021
9. Nithya Sree, B., Vipin Raj, C., Madhavan, R.: Obstacle avoidance for UAVs used in road accident monitoring. In: 2017 1st International Conference on Electronics, Materials Engineering and Nano-Technology (IEMENTech). IEEE, Kolkata (2017)
10. Fonseca, L.M.G., Namikawa, L.M., Castejon, E.F.: Digital image processing in remote sensing. In: Brazalian Symposium on Computer Graphics and Image Processing, vol. XXII, pp. 59–71 (2009)
11. Tripathi, S., Gupta, S., Kumar, V., Tiwari, P.: Hybrid utility poles and their application in power system: refinement in construction and design of conventional utility pole. In: International Conference on Electrical and Electronics Engineering (ICE3) 2020, pp. 606–611 (2020)
12. Muhammad Hanif, S., Prevost, L.: Text detection and localization in complex scene images using constrained AdaBoost algorithm. In: 2009 10th International Conference on Document Analysis and Recognition. IEEE, Barcelona (2009). https://doi.org/10.1109/ICDAR.2009.172
13. Diop, A.K.: LDA based on real-time classification of CCTV systems using codeblocks information. In: 2020 5th International Conference on Communication, Image and Signal Processing (CCISP). IEEE, Chengdu (2020)

Industrial IoT Solution for Powering Smart Manufacturing Production Lines

Ayman M. Mansour[1]([✉]), Mohammad A. Obeidat[2], and Yazan A. Yousef[3]

[1] Department of Communication, Electronics, and Computer Engineering,
College of Engineering, Tafila Technical University, Tafila 66110, Jordan
mansour@ttu.edu.jo
[2] Department of Electrical Engineering, Faculty Engineering, Al-Ahliyya Amman University,
Amman 19328, Jordan
m.obeidat@ammanu.edu.jo
[3] Ejada Industrial Company, Sahib Industrial City, Sahab, Jordan

Abstract. Iot has great benefits in all areas of life, such as factories, transportation, and airports. In this paper, a solution for monitoring the machines in a wet wipes papers factory is proposed. This factory face a problem in controlling the machines outside the working hours. One of the problems is dissolving the glue using to seal the wet wipes paper container by the heater machine about one hour before the workers come to the factory in the morning. To solve this problem, A system can be used to monitor the heater machine and this machine can be controlled online using IOT technology. The IOT can be used for both monitoring and controlling the whole process for all machines in the factory.

Keywords: Internet of Things · Factory automation · Glue heater

1 Introduction

The Internet of Things (IoT) is a system that allows devices to be connected and remotely monitored across the Internet. In the last years, the IoT concept has had a strong evolution, being currently used in various domains such as smart homes, telemedicine, industrial environments, etc. [1–5].

The production line for the manufacture of wet wipes (Fig. 1) consists of three stages. The first stage is the basic stage so that the function of the machine at this stage is to draw Tissue rolls, which are called (Nonofen), and sometimes consist of 6 or 12 rolls. So that its main function is to bend and cut the napkins according to the required size and dimensions.

Within this stage, the wipes are moistened with perfume or a moisturizing substance to obtain the required moisture percentage. After this process, through a number of transmissions, the tissues that were cut and moistened are transferred to the star, which places the tissues on top of each other in order to reach the required number and the number of tissues varies. It is called a start because of its shape and movement which is

Fig. 1. Wet wipe production line.

similar to a star. Inside the product, according to demand, it comes with 120, 72 or 52 napkins.

There are some main parts of this machine. The main motor pulls the rolls into the machine. The folds (which bend the napkin to the required width and control the width is manual, and as for the length of the napkin, it is controlled by entering the required dimension scale into the control screen only. Humidification pump, which pumps a percentage of the perfume or water during the withdrawal process, and the humidification percentage is controlled through the main control screen of the machine (machine).

The blades, which play an important role in the quality of the product, which cut the required length so that its principle is a rotational principle, that is, through an electric motor at a very high speed, which cuts the napkins, and the number of blades is 2 according to the rotational speed. The star, which are cut to certain dimensions, are emptied through transmissions and transmissions are belts with circular sections, and the material from which they are made is rubber. The star rotates at a certain speed, transferring the tissues to the elevator. The elevator principle of its work is based on the time delay, and the tissues are connected to a machine, and the napkins are arranged in it on top of each other vertically to obtain the required number of napkins either 120, 72 or 52.

The second stage, which is called the stage of packaging and welding. This machine consists of the nylon material with which the product is wrapped. The nylon is welded longitudinally and transversely, and then each piece is separated from the other (each product from the other). This machine consists of longitudinal welding discs that pull the nylon in conjunction with the entry of wet wipes inside it. And then, a knife that cuts each product from the other. At this stage, the production and expiry date is set and the batch number is called the batch number and batch number.

Temperature plays an important role in this stage, so that for each nylon, which encases the product, a certain temperature is heated through the screen for this machine, and through electric heaters, the heat called the heater is generated. The temperature is known from the thermo cable. The synchronization of the entry of wet wipes into the nylon is controlled by sensors that work on taking a reading through the wipes and thus giving an automatic order to electric motor so that it weaves the transmissions to enter the wet wipes into the nylon to complete the welding process and cut each product from the other. Note between each machine (stage) there are transmissions that move the product from one stage to another. After the second stage, the product is transferred to the third machine so that the product remains only to add the plastic cover.

The third stage is the focus of this paper. At this stage, the machine is No. 3, which gives the product the final requirements in order for the product to be usable. The plastic cover is added and the product brand is added through the so-called stickers (stamps) as shown in Fig. 2.

Fig. 2. Wet wipe cover packet fixing device.

The plastic cover is fixed with reusable glue to keep wet wipes from drying out, which is the main focus. The glue needs a temperature of 165 °C to turn from a solid state to a liquid state. So that heating is done by a heater (electrical heaters) (Fig. 3) and it takes 50 min to turn into a liquid state, in addition to the large amount in the space designated for the glue.

What is required is to make a specific program in order for the glue machine to work through an online application, so that the time required for the machine to be ready to work before starting the factory's work in the early morning is reduced.

Fig. 3. Glue heater.

2 System Architecture and Design

This section illustrates system architecture and design as well as the software configuration, The components of the real smart network are: Tow computers with network
interface card (NIC), Ethernet cable cat5 with Rj45 connecters in both side And an
Ethernet switch. Two computers have been used here to make the identification process
more realistic, the data are being transferred from one to the other and then received
again.

OPC is a compatibility framework for transferring data securely and reliably in the
industrial automation and other industries. It is platform that irrelevant and ensures a
smooth flow of data between devices from various manufacturers., OPC., it based on the
client server model as shown in Fig. 4.

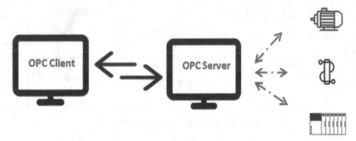

Fig. 4. OPC server.

The OPC software used in this paper is KEPSERVER, this platform's design enables
users to connect, manage, monitor, and control various automation devices and software
applications via a single user interface.

Each one of the computers must be given a specific IP address through the Ethernet switch, one of the computers will be the OPC server "KRPSERVER" and the other computer will be the OPC client "MATLAB" Fig. 5 below shows the hardware and software topology.

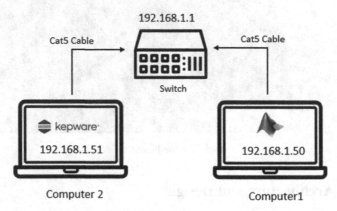

Fig. 5. Network architecture.

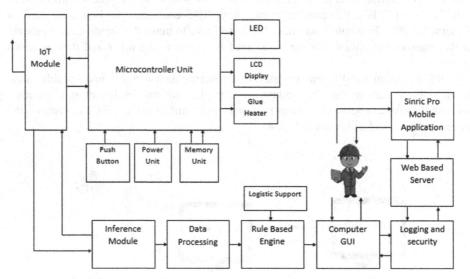

Fig. 6. System block diagram.

As you can see from Fig. 5, both computers must be in same subnet mask OPC Client - computer 1 is configured with "192.168.1.50" IP address and OPC Server – computer 2 is configured with "192.168.1.51" IP address. The developed system consists of IoT Kit, microcontroller unit, GSM module, LCD display, LEDs and glue heater unit as shown in Fig. 6.

3   The Developed IoT System

In this section, we describe the proposed IOT system. The proposed system consists of a collection of hardware and software, as illustrated in Fig. 1. The hardware module includes a microcontroller (Arduino Uno), which is connected to a relay set of type Songle Relay (SRD) as shown in Fig. 7. The relay acts as a switch. It can switch to high voltage using low power circuits. The SRD relay has three pins (IN, GND and VCC).

Fig. 7. SRD relay.

The "IN" pin of the first relay is connected to a pin on the microcontroller to switch the relay on/off and thus turn the glue heater on/off with respect to the a command either from amazon Alexa or mobile app. All appliances are working on 220 voltage.

A Java code is developed to communicate with the serial port to which the IOT is connected (Fig. 8).

Fig. 8. IoT kit.

In our experiment, we have the following sound command ("Heater on," "Heater off,"). The IoT kit received the voice command from amazon alexa (Fig. 9) or from mobile app and then it sends and order to the serial port to which the IOT is connected.

Fig. 9. Amazon Alexa.

The IOT Kit is an easy to deal with in both hardware and software. It contains everything needed to support the work of microcontroller. It can work as stand alone device or it can communicate with other software in PCs through a USB cable.

The IOT kit is programmed using the Arduino Development environment simply through connecting the board to computer via USB. The programming is considered easy task with Arduino compared with other devices that typically need an external programmer. IOT Kit receives input from the environment through a variety of sensors. The received data will be processed through the microcontroller unit and the output can affect the surrounding by controlling Buzzers, LCD, LEDS and other devices s. Arduino is an open-source electronics prototyping platform.

A prototype design for Monitoring System is successfully developed as shown in Fig. 10. This prototype design is responsible for receiving data from Alexa or mobile app. The system will be connected to PLC Unit shown in Fig. 11.

A mobile application has been developed using Sinric Pro to send commands to glue heater. The mobile application GUI is shown in Fig. 12.

The full system GUI that shows all the operation conditions of the heater and the instant temperature is shown in Fig. 13.

The following is the pseudo-code for the proposed system that is developed to make IoT performs as designed with connection ability to Alexa and Sinric Pro Mobile App. It explains how the proposed system controls the glue heater. This system provides quick response rate than the manual methods. Such systems will increase the productivity of the production line.

Fig. 10. System prototype.

Fig. 11. PLC unit.

Fig.12. Mobile App.

Fig. 13. System GUI.

```
struct RelayInfo {
  String deviceId;
  String name;
  int pin;
};

std::vector<RelayInfo> relays = {
    {"60c9dbe18cf8a303b939f888", "Heater", D1},
    {"60c9dbf88cf8a303b939f88a", "Device2", D2},
    {"60c9dc0d8cf8a303b939f88c", "Device3", D3}};

bool onPowerState(const String &deviceId, bool &state) {
  for (auto &relay : relays) {                           // for each relay configuration

    if (deviceId == relay.deviceId) {                    // check if deviceId
matches
      Serial.printf("Device %s turned %s\r\n", relay.name.c_str(), state ? "on" : "off"); // print
relay name and state to serial
      digitalWrite(relay.pin, !state);                   // set state to digital pin
/ gpio
      return true;                                       // return with success true
    }
  }
  return false; // if no relay configuration was found, return false
}
void setupRelayPins() {
  for (auto &relay : relays) {   // for each relay configuration
    pinMode(relay.pin, OUTPUT);   // set pinMode to OUTPUT
  }
  pinMode(D4, OUTPUT);
  digitalWrite(D4,LOW);
}

void setupWiFi() {
  Serial.printf("\r\n[Wifi]: Connecting");
  WiFi.begin(WIFI_SSID, WIFI_PASS);
  while (WiFi.status() != WL_CONNECTED) {
    Serial.printf(".");
    delay(250);
  }
  Serial.printf("connected!\r\n[WiFi]:        IP-Address        is        %s\r\n",
WiFi.localIP().toString().c_str());
}
void setupSinricPro() {
  for (auto &relay : relays) {                 // for each relay configuration
    SinricProSwitch &mySwitch = SinricPro[relay.deviceId];   // create a new device with
deviceId from relay configuration
    mySwitch.onPowerState(onPowerState);         // attach onPowerState callback to
the new device
  }
  SinricPro.onConnected([]() { Serial.printf("Connected to SinricPro\r\n"); });
  SinricPro.onDisconnected([]() { Serial.printf("Disconnected from SinricPro\r\n"); });
  SinricPro.begin(APP_KEY, APP_SECRET);
}

void setup() {
  Serial.begin(BAUD_RATE);
  setupRelayPins();
  setupWiFi();
  setupSinricPro();
}

void loop() {
  SinricPro.handle();
}
```

4 Conclusion

IoT system is used to monitor machines in the wet wipes paper factory and operating the machines remotely. Where sensors are installed on all the machines in the factory to monitor the factory and the entire production line. In this paper, IoT is used to solve the problem of melting glue using two ways, the mobile application or sending message online to the machine. The machine is controlled remotely by turning it on or off. IoT system can be applied in many areas, because of its great benefits in solving problem, that appear suddenly or outside working hours. The developed system saves time and efforts and contributes in increasing profit and maintaining property.

References

1. Garg, H., Dave, M.: Securing IoT devices and securely connecting the dots using REST API and middleware. In: 2019 4th International Conference on Internet of Things: Smart Innovation and Usages (IoT-SIU), Ghaziabad, India, pp. 1–6 (2019)
2. Gupta, A.K., Johari, R.: IOT based electrical device surveillance and control system. In: 2019 4th International Conference on Internet of Things: Smart Innovation and Usages (IoT-SIU), Ghaziabad, India, pp. 1–5 (2019)
3. Sharma, S., Das, S., Virmani, J., Sharma, M., Singh, S., Das, A.: IoT based dipstick type engine oil level and impurities monitoring system: a portable online spectrophotometer. In: 2019 4th International Conference on Internet of Things: Smart Innovation and Usages (IoT-SIU), Ghaziabad, India, pp. 1–4 (2019)
4. Saxena, A., Shinghal, K., Misra, R., Agarwal, A.: Automated enhanced learning system using IOT. In: 2019 4th International Conference on Internet of Things: Smart Innovation and Usages (IoT-SIU), Ghaziabad, India, pp. 1–5 (2019)
5. Vishwakarma, S.K., Upadhyaya, P., Kumari, B., Mishra, A.K.: Smart energy efficient home automation system using IoT. In: 2019 4th International Conference on Internet of Things: Smart Innovation and Usages (IoT-SIU), Ghaziabad, India, pp. 1–4 (2019)

Mobile Network Operators' Cost Efficient 5G Deployment with Hyperscalers Co-investment in Public Cloud

Gábor Földes[1,2]([✉])

[1] Finance Department, Faculty of Economic and Social Sciences,
Budapest University of Technology and Economics, Budapest, Hungary
gfoldes80@gmail.com
[2] Regulatory Department, National Media and Infocommunications Authority,
Budapest, Hungary

Abstract. The economic recovery from COVID pandemic requires a cost efficient, sustainable and green 5G rollout that should be promoted by harmonization of cost efficiency (corporate finance), competition and innovation (regulatory) aspects.

The aim of this paper to cover technology enabled new cost efficiency scenarios for Mobile Network Operators (MNO) and focus on recently emerged concern of Big Techs (Hyperscalers) co-investment into public cloud in disaggregated, virtualized RAN and Core networks.

The Research Question addressed the question of what is the optimal form to promote Hyperscalers' financial contribution to 5G rollout.

The Hypothesis tested the direct financial contribution as optimal cooperation form, claimed by many operators and politicians for 5G implementations.

The Research Methodology focused on qualitative analysis of incumbent and rival MNOs' financial incentives and competition impact of cooperation, as well as quantitative analysis applied on consultancies forecasted financial data.

The Expected finding emerged from this study is the indirect financial contribution, a co-investment partnership in public cloud might be an optimal role for Hyperscalers. The originality of this paper at academic level to highlight this is a mutual advantageous cooperation in the value-chain both for MNOs and Hyperscalers, despite having its own risk for MNOs by losing network control partially for gaining cost efficiency.

Keywords: Public cloud · Hyperscaler · Co-investment · Mobile network operator · Virtualized ran and core · Cost efficiency · Competition

1 Introduction

The current challenge of the telecom sector is the parallel rollout of capital expenditure intensive Very-High Capacity Networks (VHCN) both in mobile (5G) and fixed (FTTH) for broadband Internet, driven by predicted continuous data increase that in

F. Neri et al. (Eds.): CCCE 2022, CCIS 1630, pp. 47–61, 2022.
https://doi.org/10.1007/978-3-031-17422-3_5

case of mobile reaches CAGR 24% growth till 2027 according to latest Ericsson Mobility Report (Ericsson Mobility Report 2021). 5G small cell density and the requirement of ultra-reliability low-latency communication capabilities even further increase Total Cost of Ownership (TCO), that long-term questions the financial viability of infrastructure competition (parallel 5G networks) in all network layers. In addition, as Mobile network generations' lifecycle is becoming shorter and more overlapping eachother, the Return on Capital Employed (RoCE) figures are also worsening, many cases not reaching Weighted Average Cost of Capital (WACC), driving one of the lowest EV/EBITDA valuations for telecoms among other industries. As revenue increase is limited due to uncertain data monetization and lack-of killer applications, operational and capital expenditures (OPEX and CAPEX) have to be actively managed by operators. Therefore cost efficiency received considerable critical attention recently.

Structure. This paper has been divided into six parts. The first, introduction part deals with the research question and hypothesis, then with the explanatory drivers, as demand side of mobile network, technology enablers of cost efficiency initiatives and the savings opportunities focusing on MNO-Hyperscaler collaborations. In the second part the literature review will focus on assessment of key regulatory concerns and consultancy recommendations. The third part will cover the description of qualitative and quantitative research methodology. In the fourth, discussion part financial impacts of virtualization and co-investments will be covered, as well as show key rival and incumbent MNO strategies and incentives for Hyperscaler cooperation. The fifth, results chapter highlights the key financial impacts and the final, conclusion part presents the relevance of indirect financing via Hyperscaler co-investment model, and its potential attribution to cooperation in physical infrastructure.

Research Question. The research question seeks to address the question of what is the optimal form to promote Hyperscalers' financial contribution to 5G rollout? Clarifying that the role of Hyperscalers is gaining more importance, as their business growing by much higher pace, than MNOs business and their content-services have more powerful impact on whole value-chain financials.
The Hypothesis tested is the direct financial contribution, perceived an optimal cooperation form, claimed by many operators and politicians.
In order to make a full assessment on research question the following preconditions have to be taken into consideration.

Demand Side. Investments into more technology and spectrum efficient mobile network generations (in particular move from 4G to 5G) is already driven by customer demand, represented by traffic increase nowadays. Ericsson states, that the average mobile data consumption reached 11.4 GB/month/mobile users in 2021 and predicts CAGR 24% growth from 2021 till 2027. MNOs were struggling to identify killer applications for decades, however Over-The-Top (OTT) players, named BigTechs or Hypersclares recently managed to provide attractive content and services. Based on Sandvine reports the global internet traffic (includes both mobile and fixed, latter carries dominant part of traffic) is driven by video, gaming and social media upto 80%. Latest report published in May 2021, displayed the traffic in application category and even at application level (Cantor and Cullen 2021). Table 1 illustrates that 96% percentage of downstream traffic related to TOP6 application categories that is a really high concentration. In 2020 lockdown during the worldwide stay at home, YouTube generated 15%, Netflix 11% of total global traffic (Table 2).

Table 1. Global application category traffic share.

Ranking	Application Category	Downstream	Upstream
1	**Video Streaming**	49%	19%
2	**Social Networking**	19%	17%
3	**Web**	13%	23%
4	**Messaging**	7%	20%
5	**Gaming**	4%	2%
6	**Marketplace**	4%	2%

Table 2. Global application traffic share.

Ranking	Application	Downstream	Upstream
1	**YouTube**	20%	4%
2	**Facebook Video**	11%	3%
3	**TikTok**	7%	3%
4	**Facebook**	6%	2%
5	**Google**	5%	1%
6	**Instagram**	5%	1%

These application categories and applications are triggered by Hyperscalers, who are monetizing indeed the mobile network without contributing to the investment burden.

The customer demand has three key-attributes that will gain importance first in business then in consumer customer segments: (a), capacity, described by enhanced mobile broadband (eMBB) use cases, (b) reliability, the ultra-reliable low-latency communications (uRLLC) capability use cases, (c) capability to handle large amount of tools, the massive machine-type communication (mMTC) use cases.

Overall from demand side these factors have the largest impact on technology development directions.

Technology Enablers. What we see in technology development is far more that we got used to in previous network swaps or introduction of 4G top on 3G or even 3G top on 2G. A technology shift is going to happen towards creative destruction (disruptive technologies) by network disaggregation, virtualization and open RAN. The "monolithic telecom infrastructure based on proprietary hardware and closed interfaces" (Taga and

Virag 2021) turn to disaggregated (HW-SW) and open (multi-vendor selection) Radio Access Network (RAN) and Core Network. The legacy network failed to provide the flexibility, scalability and degree of automation. The key element of the new direction is virtualization, cloudification, softwerization that enables end to end software-defined networks (SDN), virtual network functions (VNF), cloud native, cloud edge computing, public-cloud scale, artificial intelligence (AI) driven automation and machine learning (ML).

Fields of HW & SW disaggregation, network function virtualization and containerization allow for radical architectural changes across mobile network RAN and Core domains. Virtualized network aims to put the functionality of customized equipment into software programs that run on commercial off-the-shelf (COTS) servers. In RAN domain radio signal processing can be done in the cloud. Base stations are smaller and simpler than those deployed in traditional networks, because there is no need for specialized hardware to process voice calls and data requests. Instead, the base stations simply send signals from individual devices to software in the cloud for processing. SDN and VNF allow for the scalability and automation required for future 5G use cases. Edge and far-edge data centers (DC) allow the deployment of VNFs close to the customer, reducing latency, improving quality of experience (QoE).

AI role to optimize the use of radio spectrum, including by training algorithms which are able to adapt parts of the network to specific conditions, like a large storm or concert and web-scale refers to a flexible service for robustness that can scale and add new services quickly, like a public cloud network in Core domain.

Majority of these network capabilities require to exceed currently widespread 5G non- standalone (NSA) RAN and Core networks. 5G standalone (SA) architecture is the enabler of network slicing, which is a kind of "on user-demand logical separated network partitioning, software-defined on-top of our common physical network infrastructure" (Kyllesbech 2022).

Ericsson and Arthur D. Little consultancies published a study on network slicing opportunities and revenue potential for MNOs and one of the main findings was that listed demands (eMBB, URLLC, mMTTC standard Slice Service Types, defined by GSMA and ITU) can not be served without network slicing in more than 30% of 5G use cases explored. "The diversity of requirements will only grow more disparate between use cases—the one-size-fits all approach to wireless connectivity will no longer suffice." (Network slicing…, 2021).

Overall, the innovations in RAN and Core domains indicates a reconfiguration of the existing market structure and facilitates the entry of companies that are specialized in single elements. These solutions fostering vendor diversity, increasing the development speed and competition. Market entry enabled by technologies are in particular relevant for Public Cloud service Providers (PCPs), covering Hyperscalers also to make co-investments with MNOs in the 5G Communication Service Providers (CSPs) market.

Cost Efficiency and Co-investment Models. Although 5G brings technology efficiency at unit GB cost level, economic cost efficient, sustainable and green operational excellence initiatives also required from MNO side to limit TCO increase, promote higher coverage and capacity, and allow continuously more affordable consumer prices.

Technology development is the enabling factor of new and cost efficient business models, where savings coming from (i) operator-operator collaboration, like horizontal agreements (Network Sharing) and asset reconfiguration and monetization (spin-off TowerCos and FiberCos), (ii) operator-vendor relationship reassessment with virtualized and open RAN (multi-vendor model), and (iii) operator-Hyperscalers relationship set-up enabled by cloudification of Core and RAN (public cloud providers), addressing the issue that Hyperscalers driving and monetizing GB data growth, however not contributing to infrastructure investments.

In this paper I would like to cover the assessment of MNO-Hyperscaler cost efficiency initiative from corporate finance and regulatory aspects.

Hyperscalers are those companies that offer networking, cloud and internet services with an established data center footprint. In the context of this study it primarily referring to public cloud providers. Largest players are Amazon Web Service (AWS), Microsft Azure and Google Cloud, subsidiaries of Big Tech conglomerates, owning much higher corporate valuations and revenue growth rates, than MNOs. Also significant public cloud service providers are Oracle, IBM, Alibaba, Baidu and Tencent Clouds. Hyperscalers have already for public cloud the network-centric tools (e.g.: edge computing, AI/ML analytics) and services. Due to these capabilities, MNOs consider partnering with Hyperscalers to improve cost efficiency both in CAPEX and OPEX.

2 Literature Review

A large and growing body of literature has investigated the cooperative investment (co-investment) topic, however there is a significant decrease on journal articles from fixed towards mobile, in particular to Hyperscaler subcategory.

The majority of previous studies deal with Fiber to the Home (FTTH) co-investment impact on fixed broadband internet rollout. The study of Aimene L. et al. carried out that in the French market the co-investment partnership of Orange incumbent operator and alternative providers could increase the FTTH adoption and also promote competition by lowering incumbent operator's market share. (Aimene et al. 2021).

A lower number of studies postulated co-investment in mobile segment. A recent study has argued on co-investment benefits, as infrastructure sharing agreements can decrease prices, increase consumer surplus and subscriptions (Jeanjean 2021). Another study revealed from competition policy aspects that mobile network sharing related potential harms are not jeopardizing the competition at retail market (Pápai et al. 2020).

Surprisingly very few journals conduct on Hyperscaler related co-investment opportunities. An ITU study indicated how far OTTs contribute to MNO revenue, what is the optimal form of OTTs and network operators co-investing, as data traffic accounts for a significant share of network costs. (Kettani et al. 2020).

Another study discussed the role of content providers in ultra-fast broadband access networks. It starts from network neutrality (NN) regulation of the Internet, and the question is how an Internet Service Provider (ISP) can negotiate with the content provider (CP) a fee for (priority) delivery of content and affects firms' investment incentives. The study shows that the CP might be more willing to co-invest if NN regulation were removed to avoid high ex post fees. (D'Annunzio and Reverberi 2016).

The originality of this paper that it is among the first papers at journal level shows the significance of Hyperscaler co-investment. In consultancy and business materials considerable amount of literature has been published related to concerns of Hyperscalers' market power and the necessity of regulation. However, there are relatively few studies available in the area how to resolve this interest of conflict and populate Hyperscalers into the mobile service value chain to financially contribute to 5G deployment, relieving investment burden from MNOs.

The two different approaches represented best by the appeals of regulators and recommendations of consultancies. At the section end MNOs' standpoints also introduced.

Regulation. European competition and sector regulator authorities raise concerns against mainly US based Hyperscalers.

Competition Regulation. Margrethe Vestager, EU Commissioner for Competition urged European regulators and legislators on a Financial Times conference to finalize Digital Markets Act (DMA) and the Digital Services Act (DSA) in order to curb the power of Big Techs (Espinoza 2021).

The purpose of DMA to force Hyperscalers to ensure more equal terms on their online platforms. It would affect companies with a market capitalization of at least 80bn EUR (e.g. Google, Amazon, Apple, Facebook and Microsoft) and prohibit Big Techs from ranking its own services beyond their rivals. The DSA created to clarify the way large online companies should keep illegal content off their platforms.

Summing up, the competition policy related aspects covering rather the content relevant parts and not addressing direct strategic and financial concerns.

Sector Regulation. The Body of European Regulators for Electronic Communication (BEREC) started the investigation of openRAN topic in wireless network evolution workgroup. The first published material drew 4 scenarios for most realistic developments of the 5G equipment and services supply market till 2030 (5G supply..., 2021). The study analyzed economic, technological, environmental, and social impacts for each scenario, covering key European Commission and stakeholder concerns, including market competition, costs (OPEX & CAPEX) requirements. The identified scenarios are (i) incumbent players driving 5G, (ii) slow pace of 5G rollout, (iii) open RAN as a game changer, (iiii) 5G for Big Techs.

Out of 4 scenarios, the Big Techs is relevant for this paper. Assumed storyline is that Big Techs conquer the 5G supply markets with open RAN business models, as virtualization of networks and disaggregation of software change the landscape. As software can be run on someone else's physical infrastructure, the use of clouds becomes more commonplace. Companies from vertical industries are encouraged to enter the 5G supply market. MNOs are not able to find their role as infrastructure providers for industrial players and might be overcome by Big Tech companies who also offer services to end-users. Big Techs become the "new operators" by offering carrier-like services bypassing incumbent mobile network operators and end-user connections. Thus, Big Tech companies will serve as virtual operators and increase their overall dominance in the market. Summing-up Big Tech companies may leverage their cloud and software capabilities to move into connectivity and supply domains with innovative solutions and their financial strength allows them to overtake existing players and new entrants. For Big Techs the Core network might be more relevant at first, than the RAN. Because the it is easier centralized, and closer to their current skills and infrastructure, based on cloud computing in data centers.

From Market competition perspective, there is a risk that value-add will shift from connectivity to the cloud, and MNOs' value generation may decline. Hyperscalers dominate the end-user services by offering more value-added OTT services, and MNOs remain stuck offering the least profitable part of the chain (i.e. "dumb" pipelines). Finally, Big Techs may enter into the connectivity provision market and increase competition.

Regarding cost efficiency, the study suggests that the impact on RAN-related OPEX is low. Hyperscalers take over a small part of a base station's processing, and reserving higher bargaining power and overall dominance to keep their solutions's prices high.

From US Federal Communications Commission (FCC) side a Commissioner argued, that "Big Tech has been enjoying a free ride on our internet infrastructure while skipping out in costs needed to maintain and build that network". (Carr 2021).

This statement is in line with European concerns, however formulated on a more straightforward way.

Overall the regulatory studies and statements highlighted the concerns regarding Hyperscalres entrance into telecommunication market, and put much less emphasis on potential upsides of MNO-Hyperscaler cooperation.

Consultancies. Leading technology, media and telecom (TMT) sector consultancies, like Analysys Mason, Arthur D. Little and Dell'Oro Group embrace the idea of MNO-Hyperscaler cooperation is value generating and lucrative for both parties, also addressing the question how to involve Hyperscalers into funding of 5G implementation.

Recent Analysys Mason (AM) study has conducted that MNOs should consider partnering with Hyperscalers to improve CAPEX efficiency (Brown 2021). AM suggests to set-up co-investment partnership to relieve CAPEX pressure of 5G rollout, in particular till SA rollout arrives. It also covers mainly the same co-investment models, listed in this study previously and focus on Hyperscalers cooperation scenario. AM concerns that MNOs can use third-party cloud infrastructure to support cloud-native networks and reduce CAPEX associated with building their own (telco) cloud, as Hyperscalers already have the expertise on such NW capabilities like, SDN, VNF AI/ML, edge computing and data centers. However, Brown drew attention, that "Operators must balance the benefits of Hyperscaler partnerships with the risk of losing control of the network". Overall AM recommends MNOs to assess the entry of Hyperscalers into the telecoms as an opportunity to improve CAPEX efficiency and network capabilities, as well as scale up their new networks more rapidly than they could otherwise do alone (telco cloud). Mainly Hyperscalers as new investors will not seek to challenge established MNOs in their core business (at least currently), but will be targeting a role in a value chain, transformed by cloudification.

Athur D. Little (ADL) has published a study, emphasizing the importance of virtualizing mobile networks (Taga and Virag 2021). The material offers an in-depth insight into open RAN radical architectural changes across RAN and Core domains, containing disaggregation, centralization, virtualization with multivendor model. It demonstrates the Time to Market (TTM) and customer experience (CEX) impact of Open RAN reducing lead time from weeks to hours/minutes with zero-touch operation. It also found upto

30–40% cost efficiency improvement in both OPEX and CAPEX, however these findings related to the whole ecosystem change, while Hyperscaler and MNO cooperation is just one part of it.

Dell'Oro Group draws the distinction between private (telco) and public cloud in order to demonstrate ideal partnership between MNOs and Hyperscalers (see Fig. 1, Bolan 2021b).

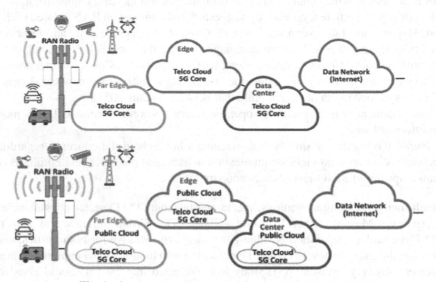

Fig. 1. Overview on Telco vs Public cloud (Bolan 2021b).

The importance of the study to explore several cooperation forms of MNO-Hyperscaler co-investments. MNOs can develop their own telco cloud however, it might be very costly to set-up cloud functionalities from scratch. As 5G networks move into the cloud, it becomes viable for Public Cloud SPs to host the 5G CSPs networks and it extends many benefits to the MNOs beyond lowering costs. Integration can be carried out via either Public Cloud SP Services go into 5G CSP Networks or Public Cloud SPs host 5G CSP Networks. Further importance of this study the listing of Public Cloud Computing advantages in terms of cost, speed, scalability, performance and productivity (Bolan 2021a).

MNO Statement. Thirteen European MNOs, member of European Telecommunications Network operators' association (ETNO) urged action from policymakers to provide a competition policy allowing operators to build the necessary scale to realize the EU's digital decade 2030 ambitions, so Europe's competition policy need to balance between investment as well as competition. (Joint CEO…, 2021).

The CEO statement raised the question on how to balance the interests of Big Techs and MNOs. Global tech players "generate and monetize" a "large and increasing part" of network traffic, which "requires continuous, intensive network investment and planning by the telecommunications sector". This can "can only be sustainable if such big tech platforms also contribute fairly to network costs", concluded the statement.

In US, similar argumentations were taken by AT&T and other ISPs over the past 15 years, that Big Techs delivering content over the Internet get a "free" ride and should subsidize the cost of building last-mile networks in Fixed line segment.

One more relevant concern raised by BT that argues for modification of net neutrality rules that stipulate the treatment of all Internet traffic equally. This rule might be considered outdated for the streaming area or 5G SA network slicing attempts, in which user-demand logic based service level combinations (differentiation service level for different prices, e.g. Latency, reliability, availability, coverage) will be offered.

Overall these studies highlight the need for conflict resolution between MNOs and Hyperscalers. Regulators revealed fairly the imbedded risk of Big Techs, that have to be mitigated and MNOs claim is valid that Hyerpscalers should contribute to network developments, but the cloudifications brings the proper, market conform role, where Hyperscalers can step into the value chain as a co-investing partner.

3 Research Methodology

The current study puts the "creative destruction" into the centre of management theories applied. Creative destruction is most often used to describe disruptive technologies such as openRAN with disaggregation, virtualization and cloudification with edge-computing that in-depth change the telecommunication landscape.

Josef Schumpeter in 1942 in Capitalism, Socialism, and Democracy book characterized creative destruction as innovations in the manufacturing process that increase productivity. "The opening up of new markets, foreign or domestic, and the organizational development from the craft shop and factory to such concerns"- illustrate the same process of industrial mutation – „that incessantly revolutionizes the economic structure from within, incessantly destroying the old one, incessantly creating a new one. This process of creative destruction is the essential fact about capitalism." (Schumpeter 2011).

Summing up, from economics point of view creative destruction describers the upcoming technological shift, where many players, including MNOs and Hyperscalers are repositioning their role.

This study employs a mixed methods approach, built on both qualitative and quantitative empirical data analysis. With regard to the research methods, some limitations need to be acknowledged, as this study covers such forward-looking, pioneer topic, in which application of actual data are limited, not only for openRAN, but in particular for public cloud economics.

Therefore, the main qualitative research method adopted to gain detailed understanding of key stakeholders' (MNOs, Hyperscalers and Regulators, covered already in Literature review chapter) market behaviour.

The limited quantitative empirical analysis can be carried-out on small size of dataset, mainly based on consultancies and vendors prognoses. Even though the relatively limited sample dataset, this study offers valuable insights into financial and competition assessment relevant considerations of MNOs and Hyperscalers cooperation.

4   Discussion

This chapter provides in-depth analysis of the market developments. The aim to present key forecasts on virtual and open RAN widespread and within trend, report public cloud evolution and the key cooperation between Hypersclaers and MNOs, divided them into challenger and incumbent operators.

Market Trend of Open RAN. Analysys Mason consultancy estimates as a mid-case (middle income countries, moderate take-up) for open RAN adoption of the market, that 54% of subscribers will be served by open RAN networks at the end of 2030. In case of slow take-up scenario 26%, in case of fast take -up scenario 86% is the relevant subscriber rate. In low income countries the growth can be a little been even higher due to higher openness for cost efficient solutions. (Abecassis et al. 2021). At investment side Rethink Technology research estimates that 58% of total RAN CAPEX and 65% of deployed sites could be openRAN related in 2026. (Everything…, 2021).

Market Trend of Communication and Public Cloud Service Providers Partnership. Over 34 partnerships have been established between MNOs and Hyperscalers as of July 2021 based on Analysys Mason to set-up a co-investment in public cloud for virtualized RAN or Core. (Brown 2021).

Challenger Communication Service Providers (MNOs). Challenger operators mainly new entrants into the matured market, who would like grab a remarkable market share. Although they might have an advantage not to build out 2-3-4G, just focus on 5G and new technologies, they suffer from front investment burden and lack of scale in short and mid-run operation. Nobuyuki Uchida, an executive at Rakuten's mobile unit told that "We came to the conclusion that we wouldn't be able to compete if we operated in the same way as the rest of the industry." (Fujikawa 2021), therefore disruptive technologies (creative destruction) like disaggregated and virtualized open RAN or MNO-Hyperscaler partnership required to increase cost efficiency (productivity) as Schumpeter stated.

Rakuten. One of the most significant trials related to Rakuten's break into the Japan market with a cloud-based wireless network. Rakuten is building Japan's fourth nationwide wireless network and has begun to offer 5G services. It virtualizes and operates through the cloud both RAN and Core parts of a mobile network. Rakuten says maintenance costs are lower in a software-driven network, because updates can be done remotely and all at once, rather than individually at base stations. Rakuten estimates 40% CAPEX savings on the cost of building out its network and 30% OPEX savings in operation. However, Rakuten rather than outsourcing to public cloud, has built its own telco cloud, and launched a subsidiary, called Rakuten Symphony, to offer the system to other operators, and via wholesale model to increase scale and achieve better return.

Dish. US based Dish planned to become a fourth national player for mobile services from previous mainly satellite-TV provider position. Dish plans to introduce a 5G network running on public cloud, operated by Amazon Web Services (AWS). Dish's network is to be the first in US that would live almost entirely in a computing public cloud, except for antennas and cables in slender boxes attached to antenna posts. These are connected directly to the AWS cloud, which hosts the virtual part of the network. Therefore Dish's fully automated network intend to be cheaper to set up and operate.

1&1 Drillisch. The German 1&1, currently Mobile Virtual Network Operator (MVNO) reported to enter into the market as fourth MNO and made a long term agreement with Rakuten to design, maintain and operate open RAN network on fully virtualized Rakuten Communication Platform (RCP).

Incumbent MNOs. Incumbent operators are adopting cloud-based technology too, but often more cautiously. The mainstream market continued to source virtualized networks from traditional suppliers in the form of pre-integrated single vendor stacks. However, the lack of interoperability between vendors' solutions created cloud-based silos.

AT&T. Many large, advanced operators, including AT&T, attempted to build their own, multi-vendor network (telco) cloud platforms, but they ran into major integration, orchestration and automation complexities and costs. (Yigit 2021).

However, in the middle of 2021 AT&T made the decision to outsource its 5G Core network to Microsoft Azure, which is a landmark decision and provides an example of a future role that Hyperscalers may take to support telecoms networks. Analysys Mason underline the significance of pioneer role of the decision: "AT&T's decision to outsource its 5G mobile network cloud to Microsoft will act as a catalyst for the adoption of public cloud provider (PCP) cloud stacks by other operators." (Yigit 2021).

AT&T will benefit from Azure's cloud expertise and scale, and will target improved QoS, as well as aiming to reduce costs. However, this partnership means that AT&T having less influence in the running of the network. This might be a CAPEX relief for AT&T to invest into other areas of its business, but it leaves the company at risk of losing control of some critical asset (Will the…, 2022).

Other MNOs. Other incumbent players also set-up co-investment partnerships with Hyperscalers to support certain goals, but mainly focusing on smaller scope, like core domain or private networks customer segment. Telefónica expanded its strategic alliance with Microsoft to deliver confidential hybrid cloud solutions to public administrations and regulated sectors. Telenor formed a partnership with Google cloud to explore how to leverage Google Cloud's expertise in data management, AI and ML and later partnered with Amazon Web Services (AWS) to speed up the modernization of its telecoms systems. Swisscom published an 5G core announcements with AWS also.

What emerges from these results reported here, MNO and Hyperscalers co-investment has started in rival MNO segment, followed by incumbents, but it is still a question, whether will collide the two huge industries and whether cloud business might jeopardise telecom industry or a balanced mutual advantageous partnership can be set-up.

5 Results

The aim of this chapter that based on key corporate financial trends, justify the necessity of cost efficient, co-investment solutions in 5G rollout and reveal the savings potential from MNO-Hyperscaler cooperation. The numeric quantitative analysis conducted on limited financial information, mainly from consultancy forecasts.

As mentioned in the literature review and discussion parts, OPEX, CAPEX, altogether Total Cost of Ownership (TCO) efficiency is required to achieve a satisfactory return on investment for 5G deployments. The following financial trends summarized from Analysys Mason consultancy and GSMA (Global System for Mobile Communications) forecasts.

OPEX. The operational cost at total telco company level globally expected slightly decrease between 2020–2027, however Network related spending forecasted to have a minor increase both nominal and proportion level (from 48% to 53%) based on Analysys Mason (Venturelli et al. 2021). The moderate increase in NW spending assumes adopting tremendous cost efficiency efforts to minimize increase from customer demand led coverage, capacity and capability investments.

CAPEX. As many consultancies, Analysys Mason also indicates that due to cost efficiency initiatives current CAPEX level can be capped.

Total CAPEX. Telecoms CAPEX will be 22% lower in 2027 than in 2019, despite new investments drop by cloud migration. (Goldman 2021) The 5G investment curve will significantly differ from 3 or 4G investment period. It will last much longer (4–6 years), without a 1–2 years peak period.

The 5G related CAPEX will reach 62% of total telco company (wireline and wireless) level in 2027 compared to 40% in 2021. (Goldman 2021).

RAN CAPEX share. GSMA estimated that RAN and Core CAPEX will slightly increase and RAN share will grow to 87% in 2025 from 82% in 2021. (Yuen 2021).

Vendor Related CAPEX Share. AM found that the CAPEX decrease is thank to decreasing MNOs' own, in-house developments and relying more on external service providers, integrators and platforms. AM argues MNOs will only achieve the cost efficiencies if they reduce their number of in-house developments as these are customized solutions that are hard to keep up-to-date and less scalable.

Vendor CAPEX share will go up to 73% in 2027 from 60% in 2017 contrast to inhouse developments. Per domain the higher increases expected in RAN to 72% from 50%, Core to 90% from 79% and IT and Cloud to 85% from 71%. (Goldman 2021).

Virtualization Share: AM study reports that despite of total CAPEX decrease expected, the virtualization relevant CAPEX domains will increase by upto 36% CAGR 2017–2027. AM prediction for increase is 36% in vRAN SW and services, 32% in vRAN cloud and radio units, 17% in SDN, soft switching and routing, 13% in cloud infrastructure, SW and edge. (Goldman 2021).

Public Cloud. The migration to public cloud from in-house developed telco cloud driven by TCO savings expectation. The adoption of cloud-based and software-centric architecture will help operators to build out significantly more network capacity than in the 4G area for the same CAPEX. MNOs end goal is "zero-touch" operations, enabling significant OPEX reductions and better real-time optimization of network performance. OPEX are lower in a software-driven network, because updates can be done remotely and all at once, rather than individually at base stations.

Therefore Analysis Mason estimated at least 25% savings potential at TCO level compared to own telco cloud development. (Gabriel and Venturelli 2022).

Summing up key financial trends conducted, the following Fig. 2. Contains the main levers.

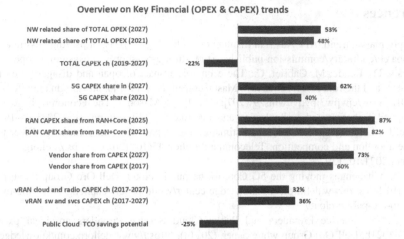

Fig. 2. Overview on Key financial (OPEX& CAPEX) trends (own statement).

6 Conclusion

Overall, the results indicate that there is a market conform, mutual beneficiary resolution for contribution of Hyperscalers to MNO investments, triggered by Hyperscalers services. Virtualization in disaggregated RAN and Core networks and Public Cloud is the direction, where telecom and Big Tech sectors get closer eachother and can set-up co-investments.

Qualitative analysis on conduct of MNOs in previous chapter demonstrates that both virtualization and public cloud are on the MNOs' agenda. Challenger telcos, striving for productivity advantage from out of-box, disruptive technologies to beat incumbents, therefore has become the pioneers in virtualization and public cloud trials. Incumbents to avoid getting into a backlog, has become opened for creative destruction to preserve their market dominance and overcome complexity in a cost efficient way, coming from their long time operation.

Quantitative analysis of consultancies' forecasted OPEX, CAPEX trends revealed that in order to limit Network TCO, cost efficiency initiatives have to be launched. Higher spending for standardized vendor solutions and virtual technologies in key domains can promote OPEX and CAPEX savings, in case of Public Cloud at least 25% TCO savings can be reached.

What emerges from this study that MNO and Hyperscaler co-investment is a feasible and viable resolution at first stage for better and more fair cooperation, however, MNOs must weigh up carefully the risks and benefits when considering partnerships with Hyperscalers. So original Hypothesis fails, direct market intervention is not necessary on the current market.

Potential Future scope might be the competition impact assessment of virtualized and open RAN with Public cloud in a network slicing capable 5G standalone network to physical infrastructure based cooperation, like Mobile Network Sharing as well as TowerCO carve-outs and mergers, that currently under strict conditions approved or even refused by National Competition Authorities.

References

5G supply market trend. AIT Austrian Institute of Technology (2021). https://digital-strategy.ec.
europa.eu/en/library/commission-publishes-study-future-5g-supply-ecosystem-europe

Abecassis, D., Kende, M. Gabriel, C.: The economic impact of open and disaggregated technologies and the role of TIP. Analysys Mason consultancy. (2021). https://cdn.brandfolder.io/
D8DI15S7/at/cjwpw3jp9qvnwmsz9w297jrx/Analysys_Mason_-_The_economic_impact_of_
open_and_disaggregated_technologies_and_the_role_of_TIP_full_report_-_2021-05-19.pdf

Aimene, L., Lebourges, M., Liang, J.: Estimating the impact of co-investment on Fiber to the
home adoption and competition. Telecommun. Policy 45(10) (2021). https://doi.org/10.1016/
j.telpol.2021.102139

Bolan, D.: Challenges moving the 5G Core tot he public cloud. Dell'Oro Group white paper.
(2021b) https://www.delloro.com/knowledge-center/white-papers/challenges-moving-the-5g-
core-to-the-public-cloud/

Bolan, D.: 5G Service Providers and Public Cloud Service Providers are ideal partners.
01.06.2021b, Dell'Oro Group white paper. (2021a). https://www.delloro.com/knowledge-cen
ter/white-papers/challenges-moving-the-5g-core-to-the-public-cloud/

Brown, E.: Operators should consider partnering with hyperscalers to improve capex efficiency.
Anal. Mason Consultancy. (2021). https://www.analysysmason.com/research/content/articles/
operator-hyperscaler-partnership-rdns0/

Cantor, L., Cullen C.: The global phenomena report. Sandvine Consul. pp. 4–
5 (2021). https://www.sandvine.com/hubfs/Sandvine_Redesign_2019/Downloads/2021/Phe
nomena/MIPR%20Q1%202021%2020210510.pdf

Carr, B.: Ending Big Tech's Free Ride. FCC. (2021). https://www.newsweek.com/ending-big-
techs-free-ride-opinion-1593696

D'Annunzio, A., Reverberi, P.: Co-investment in ultra-fast broadband access networks: is there
a role for content providers? Telecommun. Policy 40(4), 353–367 (2016). https://doi.org/10.
1016/j.telpol.2015.11.012

Ericsson Mobility report. (2021). https://www.ericsson.com/en/reports-and-papers/mobility-rep
ort/reports/november-2021

Everything you need to know about open RAN. Parallel wireless. (2021). https://www.parallelw
ireless.com/wp-content/uploads/Parallel-Wireless-e-Book-Everything-You-Need-to-Know-
about-Open-RAN.pdf

Espinoza, J.: Vestager urges European legislators to push through rules to regulate Big Tech.
Brussels, Financial Times (2021). https://www.ft.com/content/1880d0fb-0651-47ed-a8f4-6cd
e0f729859

Fujikawa, M.: To Lower Costs, Wireless Networks Such as Dish and Rakuten Head to the Cloud.
The Wall Street Journal (2021). https://www.wsj.com/articles/wireless-networks-dish-rakuten-
cloud-11634000875

Gabriel, C., Venturelli, M.: Telecoms capex and opex: moving beyond the traditional investment
model. Anal. Mason Consult. (2022). https://on24static.akamaized.net/event/35/45/80/2/rt/1/
documents/resourceList1643275400626/webinarpresentation260122final1643275399357.pdf

Goldman, L.: Cloud migration will enable vendors to increase their revenue from operators, even
while capex falls. Anal. Mason Consul. (2021). https://www.analysysmason.com/research/con
tent/articles/capex-forecast-article-rdns0/

Jeanjean, F.: Co-investment in the sharing of telecommunications infrastructures. Orange Plc.
(2021). https://papers.ssrn.com/sol3/papers.cfm?abstract_id=3934437

Joint CEO statement: Europe needs to translate its digital ambitions into concrete actions. ETNO
(2021). https://www.etno.eu/news/all-news/717-ceo-statement-2021.html

Kettani, N., Plossky, A., Hemmerlein C., Neto, G., Giovannetti, E., Martinez, J.: Economic impact of OTTs on national telecommunication/ICT markets, Geneva, Switzerland (2020). https://www.itu.int/dms_pub/itu-d/oth/07/23/D07230000030001PDFE.pdf

Kyllesbech, L.: 5G Standalone - Network Slicing, a Bigger Slice of the Value Pie (Part II). Dr. KIM, Deutsche Telekom Group (2022). https://www.linkedin.com/pulse/5g-standalone-network-slicing-slice-pie-part-ii-dr-kim/?trackingId=qPzp9akOIUdgBff%2F0D4s3w%3D%3D

Network slicing: A go-to-market guide to capture the high revenue potential. Ericsson and Arthur D. little consultancy study (2021). https://www.ericsson.com/assets/local/digital-services/network-slicing/network-slicing-value-potential.pdf

Pápai Z., Csorba G., Nagy P., Meclean A.: Competition policy issues in mobile network sharing: a European perspective. J. Eur. Compet. Law Pract. 11(7), 346–359 (2020). https://doi.org/10.1093/jeclap/lpaa018

Schumpeter, J.: Capitalism, Socialism, and Democracy. Reprint of 1947 Second Edition. Martino Fine Books. p. 83 (2011)

Taga, K., Virag, B.: Virtualizing mobile network. Arthur: Little consultancy firm. (2021). https://www.adlittle.com/en/virtualizing-mobile-networks

Venturelli, M., Gabriel, C., Brown, E.: Opex efficiency strategies for operators. Anal. Mason Consult. (2021). https://www.analysysmason.com/research/content/reports/operator-opex-efficiency-strategies-rdns0/

Will the cloud business eat the 5G telecoms industry? The Economist (2022). https://www.economist.com/business/will-the-cloud-business-eat-the-5g-telecoms-industry/21806999

Yuen, D.: Edge computing is so fun part 6: why you need to embrace the open RAN ecosystem (2021). https://hackernoon.com/edge-computing-is-so-fun-part-6-why-you-need-to-embrace-the-open-ran-ecosystem

Yigit, G.: AT&T and Microsoft Azure's deal is a major milestone for the adoption of public cloud stacks for 5G networks. Anal. Mason Consult. (2021). https://www.analysysmason.com/research/content/articles/att-azure-cloud-rma16/

Overview of Information System Testing Technology Under the "CLOUD + MIcroservices" Mode

Jianwei Zhang[1], Shan Jiang[1], Kunlong Wang[1(✉)], Rui Wang[1,2], Qi Liu[1,2], and Xiaoguang Yuan[1,3]

[1] Beijing Institute of Computer Technology and Application, Beijing 100854, China
15201122897@126.com
[2] Graduate School of the Second Research Institute of China Aerospace Science and Industry Corporation, Beijing 100854, China
[3] College of Systems Engineering, National University of Defense Technology, Changsha 410073, China

Abstract. With the continuous development of military theory, cloud computing platform and microservice technology are constantly applied to combat information platform. Under this trend, it has gradually become an urgent demand to ensure the quality of combat information system under the "Cloud + Microservice" mode. However, current studies mostly focus on a single technical point or aspect of the testing for military information system under the "Cloud + Microservice" mode, which lack of an overall summarization of its testing schema. Therefore, based on the relevant testing technologies of cloud platform, microservice and rapid environment construction in recent years, this paper summarizes and puts forward a framework for the testing of information system under the "Cloud + Microservice" mode, and systematically discusses the 3 important aspects of the framework: testing towards cloud platform, testing towards microservice applications and rapid construction of testing environment. The testing framework proposed in this paper can provide top-level guidance for the testing of such systems.

Keywords: Cloud platform testing · Microservice testing · Rapid construction of testing environment

1 Introduction

Under the guidance of some emerging military theories such as "Autonomous and Controllable", "Joint Operations" and "Network Centric Warfare", the combat center is gradually shifting from "platform" to "network", realizing the on-demand scheduling and collaborative sharing of various combat resources, giving full play to the advantages of the systematic combat, and promoting the evolution of the warfare mode from mechanization to informatization, promoting the transformation of the military command/combat system to a new generation of "network-based and service-oriented" system. At the same time, how to ensure the quality of information system under the "Cloud + Microservice" mode is becoming an urgent demand.

© The Author(s), under exclusive license to Springer Nature Switzerland AG 2022
F. Neri et al. (Eds.): CCCE 2022, CCIS 1630, pp. 62–74, 2022.
https://doi.org/10.1007/978-3-031-17422-3_6

With the promotion and use of cloud computing platform and microservice technology, research on cloud platform and microservice related test benchmarks, test frameworks and test tools has been carried out at home and abroad.

In terms of cloud platform testing, SPEC Cloud IaaS 2016 Benchmark [1] is a benchmark suite of SPEC for testing cloud performance, which aims to test the performance of IaaS layer of cloud platform; BUNGEE [2] framework is used to evaluate the elasticity of IaaS cloud platform; VMmark is used to evaluate the performance, scalability and power consumption of the virtualization platform [3]; TPC-W [4] takes the web service system as the test object, and provides testing capabilities for the scalability, peak performance, cost overhead, fault tolerance and other indicators of the cloud platform. Although there are few studies on cloud platform testing in China, relevant researchers have also done some exploratory research on cloud platform testing. He et al. of Northwest University of technology [5] have conducted a detailed study on the internal implementation mechanism of cloud storage platform. They use the "cloud test cloud" method to test the node access performance at different physical locations, which provides a new view for testing cloud storage platform; According to the general hierarchy of cloud platform, Lu et al. of Harbin University of technology put forward the method of layered testing cloud platform [6]. It can be seen that the domestic research on cloud computing platform testing is still in the stage of theoretical research and tool prototype. A complete cloud computing platform testing system and commercial tools have not been formed, and there is still a certain gap with foreign related research.

In terms of microservice testing, because of the large number of services in the microservice system, it is difficult to establish a stable test boundary, and the different characteristics of services [7] and the infrastructure used must be considered in the test stage. At present, the research on microservice testing at home and abroad focuses on component testing, end-to-end testing, regression testing and system debugging. For component testing, Camargo et al. [8] designed a test specification for single microservice testing and implemented the relevant framework; Rahman et al. [9] proposed a sandbox technology for parallel testing; For end-to-end testing, Meinke et al. [10] used machine learning method to evaluate the correctness and robustness of microservice system function during end-to-end testing; For regression testing, Kargar et al. [11] proposed an automated testing method that combines regression testing with continuous delivery of microservices; For microservice debugging, Rajagopalan et al. [12] proposed a performance fault location and repair method of microservice system.

Domestic researchers have less research on the field of microservice testing. In terms of component testing, some researchers [13] have proposed an analysis and testing method of automatically generating service dependency graph based on call chain data, so as to track and test the relationship and dependency between different services when developing and testing a new version. In terms of system debugging, Li et al. [14] proposed a microservice debugging method based on log visualization technology by using observable data. This method can jointly analyze the deployment and operation data of microservice system, so as to guide system debugging. It can be seen that foreign countries have formed relatively perfect testing strategies for the testing of microservice function, performance, correctness and robustness, while domestic research on the field of microservice testing is still relatively few. The relevant research is still in the exploratory stage, and there is a big gap between our and foreign countries on testing means and testing ability.

Although the current research has made some progress, most of them are limited to a single technical point or aspect of the testing of information system under the "Cloud + Microservice" mode, and there is a lack of combing and summarizing the overall testing scheme of information system under the "Cloud + Microservice" mode. This paper puts forward the corresponding testing technology framework for the information system in the "Cloud + Microservice" mode, expounds the three important components of the framework, and summarizes the corresponding testing tasks and related technologies of each part.

2 Test Technical Framework of Information System Under "CLOUD + Microservice" Mode

For the test of information systems in the "Cloud + Microservices" mode, a test system framework is constructed as shown in Fig. 1, which mainly includes three parts: test towards cloud platforms, test towards microservice application, and rapid construction of the test environment.

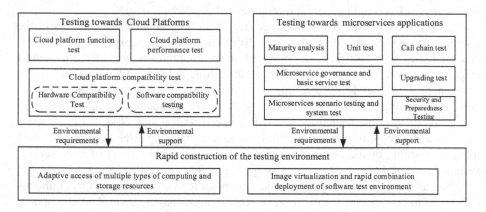

Fig. 1. Test framework for information system under "Cloud + Microservices" mode.

Test towards Cloud Platforms focuses on questions of cloud platform function, performance and basic software and hardware compatibility test;

Test towards microservice applications conducts the test of microservice applications from the perspectives of microservice maturity assessment, call relationships between microservices, microservice infrastructure, business scenarios, service upgrade, security and disaster preparedness to ensure the adequacy of the test of microservice applications.

Rapid construction of the test environment can realize the rapid combination and deployment of the basic software and hardware environments, and solve the problems of difficult in construction the test environment in the cloud environment which caused by numerous test scenarios and complex test environments.

3   Test Technologies Towards Cloud Platform

Test towards cloud platform mainly includes cloud platform function test, cloud platform performance test and compatibility test [15]. The purpose of the cloud platform function test is to ensure the correctness and integrity of the application function for each part. The function test mainly includes the test of mirror service function, host service function, monitoring service function, network service function, and storage service function [16, 17]. Performance is the core competitiveness of cloud platform [18]. The way to test and measure its performance level and improve performance is the core of cloud test, which mainly includes load balancing ability test, node failure ability test, scalability ability test and fault migration ability test. For multiple types of basic hardware and software, cloud platform compatibility test is particularly important [19]. Compatibility test generally includes hardware compatibility test and virtual machine compatibility test.

3.1   Cloud Platform Functional Test Technology

Cloud platform functional test aims to test all the functions of the cloud platform from the aspects of the correctness and integrity of the function realization by designing test cases covering the function realization of software products by using black box test technology according to the functional requirements of the cloud platform [20]. The cloud platform is a unified management platform for data center resources [21]. The main functions of cloud platform include mirror service function, host service function, monitoring service function, etc. The function test of the cloud platform also mainly focuses on cloud platform mirror service, host service, monitoring service, network service, and storage service [22]. Among them:

(1)   The test of cloud platform image service focuses on the image management function, mainly including the creation, modification, deletion, import and export of images and the management of public or private images [23, 24].
(2)   The test of cloud platform host service function mainly includes host management function test, cluster management function test and virtual resource management function test;
(3)   The test of cloud platform monitoring service function mainly tests whether the cloud platform can monitor the running status of hosts or virtual machines in the cluster and the changes of host resources [25].
(4)   The test of cloud platform network service function mainly tests whether it can manage the network resources that users have applied for, including changing, releasing, querying, configuring subnets, binding IP addresses, etc. [26];
(5)   The test of cloud platform storage service mainly tests whether users can manage the storage resources they have applied for and whether users can query, change, and release storage resources [27].

3.2   Cloud Platform Performance Test Technology

Cloud platforms have key features such as extensive network access, service measurability, rapid elasticity and scalability, on-demand self-service, resource pooling, and multi-tenancy [28]. Cloud platform performance test mainly includes load balancing capability test, node failure handling capability test, scalability capability test and fault migration capability test.

(1) Load balancing ability test [29], which mainly focuses on how to test whether the workload of the cloud platform can be evenly distributed among multiple computer nodes, processes, disks and other resources, so as to maximize the utilization rate of system resources, effectively reduce service waiting time and improve service quality.

(2) Node failure handling ability test [30], which main focuses on how to test whether cloud computing as a distributed cluster can ensure that it can provide users with stable and reliable services even when local nodes fail. In the evaluation of cloud platform node failure performance, it is usually measured by whether a virtual machine can provide services to measure whether the node is alive. In addition, the test is completed by manually manufacturing the failed node by disabling the wired network adapter on the tested host.

(3) Scalability test [31], which mainly focuses on how to test the ability of the cloud platform to maintain the service level of the platform by applying for and recovering resources from the platform virtual resource pool as the number of tenants changes and the number of applications in the platform changes.

(4) Failover ability test [32], which mainly focuses on how to test whether the cloud platform can realize the platform resource service that provides users with almost no difference when some kind of error or failure occurs in the process of using resources through virtualization, live migration and related management technologies when one or more virtual machines break down due to some reason or some fault.

3.3 Cloud Platform Compatibility Test Technology

The application architecture of information system in cloud environment is gradually changing from traditional single architecture to microservice architecture, and the deployment mode is changing from single-node to multi-node cloud environment [33]. The compatibility of cloud platform is an important criterion to measure whether the information system in cloud environment is available. The cloud platform compatibility test uses configuration or installation, and cross-platform tests technologies to test the compatibility of the cloud platform from the aspects of hardware compatibility and virtual machine compatibility by designing test cases based on the cloud platform compatibility requirements. Cloud platform compatibility test is of great significance for assessing the compatibility and availability of information systems in the cloud environment. It is generally carried out from two aspects: cloud platform hardware compatibility test technology and cloud platform software compatibility test technology.

Checking Cloud Platform Hardware Compatibility. Hardware compatibility testing mainly refers to the compatibility testing of each type of specific hardware based on the general specifications of operating system hardware compatibility and for different hardware characteristics [34], generally including CPU compatibility testing and storage device compatibility testing. The purpose of the CPU compatibility test is to ensure that the computing and control nodes of the information system cloud deployment platform can run properly. Storage device compatibility is usually checked during and after the installation is successful to verify the correct response of the operating system to the connection request of SATA [35], SAS [36] and SSD [37] type mechanical hard disk or solid state disk. Common detection items include model identification, interface and bus type identification, capacity information identification, partition creation and deletion, formatting, and read/write data. In addition, sequential read/write, random read/write performance, and read/write pressure tests are performed.

Checking Software Compatibility in the Cloud Environment. The software and hardware indicators of different operating systems are tested to determine whether they meet the basic requirements for information system installation and operation in the cloud environment [38]. Software compatibility test in the cloud environment mainly includes: (1) kernel support test. For different operating system platforms, the installation of database products under the platform is detected successively, and the test environment is formulated to check whether its functions run normally [39]. (2) Compilation environment detection. Test the corresponding development environment, such as JDK environment (whether the version meets the minimum requirements), C++ environment (GCC dependency package version), etc., according to the requirements of the compiled language in the operating environment requested by the product technical documentations [40]. (3) Dependency packet detection. Depending on the operating system (OS), the dependency packages are processed differently. If the basic database package environment is provided by the OS, checking whether the OS has all the dependency packages based on the requirements described in the product technical documentation [41].

4   Testing Techniques for Microservice Applications

Due to the distributed, dynamic and multi-service characteristics of microservice system, multiple services will run at the same time, call each other and produce dynamic changes during operation, which brings challenges to the testing of microservice system [42], such as: how to test the system under different environments/configurations, how to test the complex dependencies between services/modules, how to analyze and locate faults under complex interactions, and how to carry out the microservice system testing between different teams in parallel, etc. [43]. The testing of microservice application mainly includes [44]: microservice maturity analysis, single microservice testing, microservice invocation chain testing, microservice governance and basic service testing, microservice scenario and system testing, microservice upgrade testing, microservice security and disaster recovery testing, etc.

4.1   Microservice Maturity Analysis

The maturity of the microservice system refers to the practice level achieved by the microservice team in different aspects such as architecture, engineering practice and organizational culture [45]. By analyzing the maturity of a microservice system in different dimensions, the bottlenecks existing in the current system state and the future improvement direction can be identified to further improve the system [46].

For the maturity analysis of microservice system, it can be divided into service division, data division, testing, deployment and release, service fault tolerance and other dimensions to analyze the maturity from the static perspective [47, 48].

4.2 Single Microservice Testing

The test of single microservice mainly focuses on service interface test and service configuration test [49]. Service interface test mainly includes service function and exception handling of single interface, and service configuration test mainly includes service deployment and configuration [50], middleware or persistence layer infrastructure-related configuration. This process also corresponds to the integration test and component test in the test process [51], which verifies that a single microservice meets the design objectives, occupy the expected resources, can be independently deployed and provide the desired functions [52].

4.3 Microservice Invocation Chain Testing

Microservice call chain test refers to collect the call chain data generated during the running of microservice system by using call chain collection tools, so as to analyze the rationality of system call and dependency design and test whether the interface call results meet expectations [53]. Generally, microservice call chain test includes call chain analysis test, dependency analysis test, interface test, etc. Call chain analysis test is the key technology used in microservice fault location and abnormal early warning. By using the call chain data generated by the runtime request, it can find the potential problems in the system and make early warning in time, so as to provide more reliable system information for the development and operation staff [54]. The dependency analysis test mainly analyzes the service dependency through the call chain, so as to judge the rationality of microservices. Interface tests are designed to help validate the collaborative interactions between microservices and verify that the microservices interfaces exposed to clients or other microservices work properly.

4.4 Microservice Governance and Basic Service Testing

The proper operation of microservice systems depends on a lot of infrastructures and rational governance policies [55]. The typical basic services in microservice system include service registry, API gateway, load balancer, etc. There are often several services in a microservice system, which often leads to an avalanche effect. If one service is unavailable, cascading failures will occur, which will lead to the unavailability of the entire system [56]. Microservice governance and basic service testing determine whether the fault-tolerant capabilities of infrastructure and technology on which microservice system operation depends meet expectations by testing the capabilities of service registry, gateway, service fuse, and traffic limiting [57].

4.5 Microservice Scenario and System Testing

Microservice component in the system has the dynamic change characteristics, and microservice depends on many base configurations of system, such as network, containers, virtual machines [58]. Changes in these conditions have a great impact on microservice architecture and runtime state, so it is necessary to test microservice systems in all kinds of deployment and running scenarios, in order to ensure that the microservice system can provide the expected functions in dynamic changing scenarios [13]. Considerations include dynamic scaling, service status checking, data consistency, and microservice status when the deployment environment changes.

4.6 Microservice Upgrade Testing

Because a microservice system is deployed on cloud resources (containers or VMS) and consists of multiple services, the system upgrade is different from a single architecture [59, 60]. The following issues should be considered in microservice system upgrade: (1) how to ensure that the microservice is not interrupted during the upgrade process. (2) how to ensure that the functions of the new version microservices are available. (3) how to roll back and restore functions in case of upgrade problems. (4) whether data remains intact after upgrade and migration.

Microservice system upgrade is usually achieved through the following methods: (1) rolling upgrade [61]. Take out some service instances in the system for update, and put them back into use until all systems are updated. (2) Gray/Canary release. Maintain two versions in the upgrade process, distribute the service access traffic to different versions of services according to weights or rules [62]. (3) Blue-green deployment. Don't stop the service of the old version, deploy the new version and then test it, and switch the entire system to the new version after the test is completed.

4.7 Microservice Security and Disaster Recovery Testing

Microservice system is highly distributed because it is deployed on cloud resources (containers or virtual machines) and consists of many components. Compared with traditional software, it changes the way of software security protection [63]. Due to the characteristics of microservices, there are a large amount of distributed data traffic and environment configurations in the system, and components change dynamically, which also need to be considered in security testing. The security test of the microservice system mainly includes the following parts: (1) it is necessary to test whether the microservice interface has authentication ability. For the internal interface, front-end interface, and north-south interface in the system, whether corresponding permissions are managed respectively, and sensitive commands or system changes in the system are logged. (2) Check whether the data flows, configuration files and persistent data in the microservice system contain sensitive information or unauthorized commands, and whether the encryption and decryption framework of the system is reliable [64]. (3) Penetration test and fuzzy test should be carried out on the external interface provided by the system, in order to prevent attackers from attacking the system through injection and forgery.

5   Rapid Construction of Test Environment

The rapid construction of large-scale test environment involves the adaptation and access of multi-type computing and storage resources, and the image virtualization and rapid combined deployment of software test environment [65]. Among them, the adaptation access of multi-type computing and storage resources focuses on how to realize the unified management of heterogeneous computing and storage resources through virtualization; The image virtualization and rapid combined deployment of software test environment focus on the rapid deployment of virtual image and test environment, and provide a test environment for software testing flexibly based on test requirements.

5.1   Adaptation and Access of Multi-type Computing and Storage Resources

The adaptation and access of multi-type computing and storage resources mainly includes container based computing resource virtualization, distributed storage based storage virtualization and heterogeneous CPU computing resource fusion management [66, 67]. Container based virtualization of computing resources, virtualization and scheduling management of resources on hardware devices, unified scheduling and management of computing resources and improvement of resource utilization; Storage virtualization based on distributed storage integrates the storage resources of each computing node into a unified storage resource pool through distributed storage technology to provide storage services for the virtualization plane [68]; Heterogeneous CPU computing resource integration management technology constructs a set of middleware management platform organically combined with the hardware layer through middleware technology [69], provides standardized service interfaces for applications, shields the heterogeneity between different platforms at the bottom and the compatibility between hardware and operating system, and manages the communication, node resources and coordination between nodes in the distributed system, Provide unified services for upper layer applications.

5.2   Image Virtualization and Rapid Combined Deployment of Software Test Environment

In order to realize the image virtualization and rapid combination deployment of software test environment, we need to start from the following four aspects: first, establish the classified management and formal description of resources to realize the grouping management of heterogeneous images and test suites [70]; Second, shield low-level implementation differences and provide a unified interface for cloud platform test resource management and distribution through heavyweight virtual machine (KVM) [71] or lightweight container (docker) technology [72]; Third, based on distributed image storage and instantiation technology, realize efficient and secure image management mechanism to meet high concurrency and high throughput image requests and secure access to the test environment; Fourth, the fast combination of image and test suite based on hierarchical file system, generate the combined image required for test on demand, and support the fast combination of multi application test suite [73].

6 Conclusion

Focusing on the testing of information system in the mode of "cloud + microservice", in this paper, we present a systematic framework for information system testing under the "Cloud + Microservice" mode, which is different from the conventional research that only focus on single technical point or aspect. Based on this framework, we further introduce the related technologies and elaborates their connotation for cloud platform testing, microservice application testing and rapid construction of test environment respectively.

The testing framework proposed in this paper can provide top-level guidance for the testing of information system under the "Cloud + Microservice" mode. However, the specific implementation of such testing still needs further practice to truly ensure the quality and efficiency of the information system under the "cloud + micro service" mode through testing means.

References

1. Baset, S., Silva, M., Wakou, N.: Spec cloud™ IaaS 2016 benchmark. In: Proceedings of the 8th ACM/SPEC on International Conference on Performance Engineering, pp. 423–423 (2017)
2. Herbst, N.R., Kounev, S., Weber, A., et al.: Bungee: an elasticity benchmark for self-adaptive IAAS cloud environments. In: 2015 IEEE/ACM 10th International Symposium on Software Engineering for Adaptive and Self-Managing Systems, pp. 46–56. IEEE (2015)
3. Makhija, V., Herndon, B., Smith, P., et al.: VMmark: a scalable benchmark for virtualized systems. Technical Report TR 2006–002, VMware (2006)
4. Menascé, D.A.: TPC-W: a benchmark for e-commerce. IEEE Internet Comput. 6(3), 83–87 (2002)
5. He, Q.: The Research on Key Technology in Deduplication on Cloud Storage. Northwest University of Technology (2016)
6. Lv, D.: Research on the Perfromance Evaluation for Cloud Platform. Harbin Institute of Technology (2014)
7. Seppo, J.: Sirkemaa, information systems management – understanding modular approach. J. Adv. Inf. Technol. 10(4), 148–151 (2019). https://doi.org/10.12720/jait.10.4.148-151
8. De Camargo, A., Salvadori, I., Mello, R.S., et al.: An architecture to automate performance tests on microservices. In: Proceedings of the 18th International Conference on Information Integration and Web-Based Applications and Services, pp. 422–429 (2016)
9. Rahman, M., Chen, Z., Gao, J.: A service framework for parallel test execution on a developer's local development workstation. In: 2015 IEEE Symposium on Service-Oriented System Engineering, pp. 153–160. IEEE (2015)
10. Meinke, K., Nycander, P.: Learning-based testing of distributed microservice architectures: correctness and fault injection. In: Bianculli, D., Calinescu, R., Rumpe, B. (eds.) SEFM 2015. LNCS, vol. 9509, pp. 3–10. Springer, Heidelberg (2015). https://doi.org/10.1007/978-3-662-49224-6_1
11. Kargar, M.J., Hanifizade, A.: Automation of regression test in microservice architecture. In: 2018 4th International Conference on Web Research (ICWR), pp. 133–137. IEEE (2018)
12. Rajagopalan, S., Jamjoom, H.: App–bisect: autonomous healing for microservice-based apps. In: 7th {USENIX} Workshop on Hot Topics in Cloud Computing (HotCloud 15) (2015)
13. Shang-Pin, M.,Chen-Yuan, F., Yen, C., I-Hsiu, L., Ci-Wei, L.: Graph-based and scenario-driven microservice analysis, retrieval, and testing. Future Gener. Comput. Syst.100, 724–735 (2019)

14. Wenhai, L., Xin, P., Dan, D., et al.: Method of microservice system debugging based on log visualization analysis. Comput. Sci. **46**(11), 145–155 (2019)
15. Katherine, A.V., Alagarsamy, K.: Software testing in cloud platform: a survey. Int. J. Comput. Appl. **46**(6), 21–25 (2012)
16. Tangirala, S.: Efficient big data analytics and management through the usage of cloud. Architecture **7**(4), 302–307 (2016). https://doi.org/10.12720/jait.7.4.302-307
17. Shams, A., Sharif, H., Helfert, M.: A novel model for cloud computing analytics and measurement. J. Adv. Inf. Technol. **12**(2), 93–106 (2021). https://doi.org/10.12720/jait.12.2.93-106
18. Hazra, D., Roy, A., Midya, S., et al.: Distributed task scheduling in cloud platform: a survey. In: Satapathy, S., Bhateja, V., Das, S. (eds.) Smart computing and informatics, vol. 77, pp. 183–191. Springer, Singapore, (2018). https://doi.org/10.1007/978-981-10-5544-7_19
19. Marozzo, F.: Infrastructures for high-performance computing: cloud infrastructures (2019)
20. Bertolino, A., Angelis, G.D., Gallego, M., et al.: A systematic review on cloud testing. ACM Comput. Surv. (CSUR) **52**(5), 1–42 (2019)
21. Souri, A., Navimipour, N.J., Rahmani, A.M.: Formal verification approaches and standards in the cloud computing: a comprehensive and systematic review. Comput. Stand. Interfaces **58**, 1–22 (2018)
22. Sarabdeen, J., Ishak, M.M.M.: Impediment of privacy in the use of clouds by educational. Institutions **6**(3), 167–172 (2015). https://doi.org/10.12720/jait.6.3.167-172
23. Mohamed, S., Hadj, B.: Mobile cloud computing: security issues and considerations. **6**(4), 248–251 (2015). https://doi.org/10.12720/jait.6.4.248-251
24. Osman, G., et al.: Security measurement as a trust in cloud computing service selection and monitoring. **8**(2), 100–106 (2017). https://doi.org/10.12720/jait.8.2.100-106
25. Cornetta, G., Mateos, J., Touhafi, A., et al.: Design, simulation and testing of a cloud platform for sharing digital fabrication resources for education. J. Cloud Comput. **8**(1), 1–22 (2019)
26. Kotas, C., Naughton, T., Imam, N.: A comparison of amazon web services and microsoft azure cloud platforms for high performance computing. In: 2018 IEEE International Conference on Consumer Electronics (ICCE), pp. 1–4. IEEE (2018)
27. Arif, H., Hajjdiab, H., Al Harbi, F., et al.: A comparison between Google cloud service and iCloud. In: 2019 IEEE 4th International Conference on Computer and Communication Systems (ICCCS), pp. 337–340. IEEE (2019)
28. Mahmud, K., Usman, M.: Trust establishment and estimation in cloud services: a systematic literature review. J. Netw. Syst. Manage. **27**(2), 489–540 (2019)
29. Liu, H., Niu, Z., Wu, T., et al.: A performance evaluation method of load balancing capability in SaaS layer of cloud platform. J. Phys. Conf. Ser. **1856**(1), 012065 (2021)
30. Lin, Q., Hsieh, K., Dang, Y., et al.: Predicting node failure in cloud service systems. In: Proceedings of the 2018 26th ACM Joint Meeting on European Software Engineering Conference and Symposium on the Foundations of Software Engineering, pp. 480–490 (2018)
31. Bai, X., Li, M., Chen, B., et al.: Cloud testing tools. In: Proceedings of 2011 IEEE 6th International Symposium on Service Oriented System (SOSE), pp. 1–12. IEEE (2011)
32. Addo, I.D., Ahamed, S.I., Chu, W.C.: A reference architecture for high-availability automatic failover between PaaS cloud providers. In: 2014 International Conference on Trustworthy Systems and their Applications, pp. 14–21. IEEE (2014)
33. Burns, B.: Designing Distributed Systems: Patterns and Paradigms for Scalable, Reliable Services. O'Reilly Media, Inc., Sebastopol (2018)
34. Zhang, T., Gao, J., Cheng, J., et al.: Compatibility testing service for mobile applications. In: 2015 IEEE Symposium on Service-Oriented System Engineering, pp. 179–186. IEEE (2015)
35. Jeevitha, L., Umadevi, B., Hemavathy, M.: SATA Protocol implementation on FPGA for write protection of hard disk drive/Solid state device. In: 2019 3rd International Conference

on Electronics, Communication and Aerospace Technology (ICECA), pp. 614–617. IEEE (2019)

36. Bezrukov, I.A., Salnikov, A.I., Yakovlev, V.A., et al.: A data buffering and transmission system: a study of the performance of a disk subsystem. Instrum. Exp. Tech. **61**(4), 467–472 (2018)
37. Xu, E., Zheng, M., Qin, F., et al.: Lessons and actions: what we learned from 10k SSD-related storage system failures. In: 2019 {USENIX} Annual Technical Conference ({USENIX}{ATC} 19), pp. 961–976 (2019)
38. Yoon, I.C., Sussman, A., Memon, A., et al.: Effective and scalable software compatibility testing. In: Proceedings of the 2008 international symposium on Software Testing and Analysis, pp. 63–74 (2008)
39. Feyzi, F., Parsa, S.: Kernel-based detection of coincidentally correct test cases to improve fault localisation effectiveness. Int. J. Appl. Pattern Recogn. **5**(2), 119–136 (2018)
40. Liu, C., Yang, H., Sun, R., et al.: Swtvm: exploring the automated compilation for deep learning on sunway architecture. arXiv preprint arXiv:1904.07404 (2019)
41. Tahvili, S., Hatvani, L., Felderer, M., et al.: Automated functional dependency detection between test cases using doc2vec and clustering. In: 2019 IEEE International Conference on Artificial Intelligence Testing (AITest), pp. 19–26. IEEE (2019)
42. Feng Zhiyong, X., Yanwei, X.X., Shizhan, C.: Review on the development of microservice architecture. J. Comput. Res. Dev. **57**(5), 1103–1122 (2020). https://doi.org/10.7544/issn1000-1239.2020.20190460
43. Chun-xia, L.: Research overview of microservices architecture. Softw. Guide **18**(8), 1–3,7 (2019) https://doi.org/10.11907/rjdk.182825
44. Muhammad, W., Peng, L., Mojtaba, S., Amleto, D.S., Gastón, M.: Design, monitoring, and testing of microservices systems: the practitioners' perspective. J. Syst. Softw. **182**, 111061 (2021)
45. Jie, D.: Research on application performance test method based on microservice architecture. Digital User **27**(3), 72–75 (2021)
46. Zhou, Y., Kan, L., Peng, Z.: Design of test platform based on micro service architecture and continuous delivery technology. China Comput. Commun. **23**, 76–77 (2017)
47. Chang, Y.: Design and Development of Interface Automation Test Services and Report Generation based on Microservices Architecture. Inner Mongolia University, Inner Mongolia (2019)
48. Yuanbing, Z.: Automated testing based on microservices architecture. Electron. Technol. Softw. Eng. **4**, 119–120 (2019)
49. Huayao, W., Wenjun, D.: Research progress on the development of microservices. Comput. Res. Dev. **57**(3), 525–541 (2020). https://doi.org/10.7544/issn1000-1239.2020.20190624
50. Shengji, Q.: Brief introduction to MOCK testing technology in microservice system. Digital User **24**(23), 89 (2018). https://doi.org/10.3969/j.issn.1009-0843.2018.23.080
51. Zhou, Y., Kan, L., Peng, Z.: Analysis of software testing mode transformation under microservice architecture. Comput. Knowl. Technol. **13**(35), 83–84 (2017)
52. Chen, J., Chen, M., Pu, Y.: B/S system performance analysis based on microservices architecture. Comput. Syst. Appl. **29**(02), 233–237 (2020). https://doi.org/10.15888/j.cnki.csa.007285
53. Rahman, M., Gao, J.: A reusable automated acceptance testing architecture for microservices in behavior-driven development (2015)
54. Bento, A., Correia, J., Filipe, R., et al.: Automated analysis of distributed tracing: challenges and research directions. J. Grid Comput. **19**(1), 1–15 (2021)
55. Hao, D., Xie, T., Zhang, L., et al.: Test input reduction for result inspection to facilitate fault localization. Autom. Softw. Eng. **17**(1), 5–31 (2010)

56. Lei, Q.: Microservice performance simulation test based on Kubemark. Comput. Eng. Sci. **42**(07), 1151–1157 (2020)
57. Xuan, M.: Researchs on Microservice Invocation Based on Spring Cloud. Wuhan University of Technology, China (2018)
58. Shan, S., Marcela, R., Jiting, X., Chris, S., Nanditha, P., Russell, S.: Cost study of test automation over documentation for microservices. In: Proceedings of 2018 International Conference on Computer Science and Software Engineering (CSSE 2018), pp. 290–305 (2018)
59. Chang, Y.: Design and Development of Interface Automation Test Services and Test Report Generation Based on Microservices Architecture. Inner Mongolia University (2019)
60. Jingyi, X.U., Zeyu, Z.H.A.O., Minhu, S.H.E.N., Yibin, Y.I.N.G., Weiqiang, Z.H.O.U.: Next generation IP network test system framework based on microservices architecture. Telecom Sci. **35**(09), 29–37 (2019)
61. Tian, B., Wang, W., Su, Q., et al.: Research on application performance monitoring platform based on microservice architecture. Inf. Technol. Inf. (1), 125–128 (2018). https://doi.org/10.3969/j.issn.1672-9528.2018.01.030
62. Bi, X., Liu, Y., Chen, F.: Research and optimization of network performance of microservice application platform. Comput. Eng. **44**(5), 53–59 (2018). https://doi.org/10.19678/j.issn.1000-3428.0047130
63. Binghu, Y.: Design and implementation of mobile application security detection system based on microservice architecture. Digital Technol. Appl. **36**(11), 169–171 (2018). https://doi.org/10.19695/j.cnki.cn12-1369.2018.11.91
64. Qian, Z., Kan, L., Zhou, Y.: Analysis of mobile application compatibility test implementation of testing cloud platform based on micro-service architecture. Sci. Technol. Inf. **16**(28), 19–20 (2018). https://doi.org/10.16661/j.cnki.1672-3791.2018.28.019
65. Xing, X., Yinqiao, L., Xuefeng, L., et al.: Rapid deployment for enterprise development and test environment. Ind. Control Comput. **31**(3), 12–14 (2018)
66. Yuming, Z.: Research on Resource Collaboration and Adaption Mechanisms in Smart Identifier Networking for Edge Computing. Beijing Jiaotong University, China (2021)
67. Bo, L., Jianglong, W., Qianying, Z., et al.: Novel network virtualization architecture based on the convergence of computing, storage and transport resources. Telecomm. Sci. **36**(7), 42–54 (2020)
68. Zhengfeng, J., Keyi, Q., Meiyu, Z.: Two-stage edge service composition and scheduling method for edge computing QoE. J. Chinese Comput. Syst. **40**(07), 1397–1403 (2019)
69. Hejji, D.J., Nassif, A.B., Nasir, Q., et al.: Systematic literature review: metaheuristics-based approach for workflow scheduling in cloud. In: 2020 International Conference on Communications, Computing, Cybersecurity, and Informatics (CCCI), pp. 1–5 (2020)
70. Wei, Y., Pan, L., Liu, S., et al.: DRL-scheduling: an intelligent QoS-aware job scheduling framework for applications in clouds. IEEE Access **6**, 55112–55125 (2018)
71. Ya, L., Li, L., Xilin, Z.: Research on virtual machine static migration technology based on KVM. Sci. Technol. Innov. **25**, 85–86 (2021)
72. Kai, W., Gongxuan, Z., Xiumin, Z.: Research on virtualization technology based on container. Comput. Technol. Dev. **000**(008), 138–141 (2015)
73. Piraghaj, S.F., Dastjerdi, A.V., Calheiros, R.N., et al.: ContainerCloudSim: an environment for modeling and simulation of containers in cloud data centers. Soft. Pract. Exp. **47**(4), 505–521 (2017)

Image Denoising Using a Deep Auto-encoder Approach Based on Beetle Antennae Search Algorithm

Qian Xiang[(⊠)] and Peng Zhu

College of Information Science and Engineering, Wuchang Shouyi University,
Wuhan, People's Republic of China
xq_new21@163.com

Abstract. Image will be polluted by different kinds of noise in the process of acquisition, compression, and transmission, resulting in interference to subsequent image segmentation, feature extraction and other processing. With the development of deep convolutional neural network (DCNN), quite a few effective DCNNs have been designed and have made remarkable progress in image denoising. Gradient descent algorithm is generally used for DCNN training. However, due to the complex mathematical properties of the high-dimensional and non-convex loss optimization surface, there are often many local optimal points, saddle points or large range of gradient gentle regions, which affects training effect of the gradient descent algorithm. Although intelligent algorithms such as evolutionary algorithm have global optimization capability, they often have large computing resource requirements and slow convergence speed, which limit its application in DCNN training which is a high-dimensional optimization problem. Beetle antennae search (BAS) algorithm is a simple and efficient bionic intelligent optimization algorithm, which has global search ability. In this paper, the gradient descent method and BAS method are combined as a hybrid method for deep auto-encoder (DAE) denoising network training. Experimental results show that the proposed method accelerates the training speed of the DAE denoising network, reduces the blurring of edge details and improves the visual effect of the restored image.

Keywords: Beetle Antennae Search Algorithm · Convolutional neural network · Image denoising

1 Introduction

Image denoising aims to reduce or eliminate the interference of noise to the image and make the processed image as close to the original image as possible. As it is difficult to obtain clean-noise image training data, additive noise model is usually used to generate noise image, which can be expressed as $y = x + v$, where v is the noise. In the past decades, different kinds of image denoising algorithms have been proposed, and they can be divided into four categories according to the characteristics of denoising methods: (1) thresholding-based filtering and denoising [1, 2]; (2) sparse representation-based

© The Author(s), under exclusive license to Springer Nature Switzerland AG 2022
F. Neri et al. (Eds.): CCCE 2022, CCIS 1630, pp. 75–84, 2022.
https://doi.org/10.1007/978-3-031-17422-3_7

algorithms [3, 4], (3) partial differential equations (PDEs) based algorithms [5, 6]; (4) image self-similarity-based method [7, 8]. Although traditional image denoising methods have achieved good denoising performance, they still exist some disadvantages, especially the need of task specific model optimization for different noise reduction. For non-Gaussian image noise, it is difficult to achieve good denoising effect in complex texture area (mostly edge area), and there exists phenomena such as detail loss and blur.

With the development of deep convolutional neural network (CNN), deep convolutional neural network (DCNN) image denoising attracts more and more researchers, because it does not need to select image features and can directly realize self-learning through a large amount of image data to achieve end-to-end denoising. Compared with the traditional methods, the image denoising method based on DCNN [9–12] can extract shallow pixel level features and deep semantic level features autonomously and quickly and has strong representation learning ability and good denoising effect. DCNN can improve feature expression ability by increasing the number of convolutional layers, but complex networks also bring large number of parameters, increasing training difficulty, training is easy to fit and other problems, and traditional gradient methods are difficult to deal with such problems.

Meta-heuristic algorithms have a wide application in the field of application because of their flexibility and robustness and ability to avoid local optima. Common with particle swarm optimization algorithm (PSO) [13], artificial swarm optimization algorithm (ABC) [14], fruit flies optimization algorithm (FOA) [15], the genetic algorithm (GA) [16]. Jiang et al. [17] proposed the beetle antennae search (BAS) algorithm inspired by beetle foraging and mating behavior. Beetles pick up the smell of food and potential mates in the air and proceed according to the smell concentration detected by antennae on both sides. BAS has low complexity and does not need to know the gradient information to achieve the purpose of optimization. BAS algorithm also requires fewer initial parameters, so it is less affected by parameter sensitivity. BAS's core code is very short, only four lines, which is easy to implement and write in any language. Compared with bat algorithm (BA) and artificial bee colony (ABC), it has higher efficiency and lower complexity. In addition, because there is only one beetle during the iteration, it is lower in time and space complexity and more efficient than most swarm intelligence algorithms.

In this paper, a hybrid algorithm is proposed by combining the gradient descent optimization algorithm with BAS algorithm and is applied to deep network training. Under the premise of not increasing the computational complexity, the proposed algorithm overcomes the shortcomings of single gradient descent algorithm, such as easy convergence to the local optimal and slow convergence speed, and effectively improves the denoising performance of DAE.

2 Related Work

2.1 DCNN Based Image Denoising

Vincent et al. [19] introduced the stacked denoising auto-encoder (SDAE) and indicated that its denoising performance was comparable to K-singular value decomposition (K-SVD). Zhang et al. [20] proposed a feed-forward denoising convolutional neural networks (DnCNN) combining batch normalization and residual learning techniques and achieved outstanding denoising effects. However, this method is not robust enough to complex real-world noise as over-fitting to the additive Gaussian white noise model. To handle more complex real noise, Zhang et al. [11] presented a fast and flexible denoising convolutional neural network (FFDNet) with the ability to manage a wide range of noise levels and remove spatially variant noise by specifying an adjustable noise level map. Aiming at the difficulty on obtaining noiseless image, Moran et al. [9] proposed a method (Noiser2Noise) which use only noise samples and reach closing performance to model trained by uising paired clean-noise data. To improve the denoising performance for complex real-world noise. Guo et al. [21] designed a convolutional blind denoising network (CBDNet) for real photographs. CBDNet perform robustly when the noise model is not well matched with real-world noise. Most of these method use gradient descent to train the network. However, due to the complex mathematical properties of the high-dimensional, non-convex loss optimization surface, there are often many local optimal, saddle points or large range of gradient gentle regions, which affects the training effect of gradient descent algorithm.

2.2 Beetle Antennae Search Algorithm

The BAS algorithm has been successfully applied in many fields and has been growing, such as robot path planning and obstacle avoidance, PID parameter setting, image enhancement, fault diagnosis, neural network and so on. Cai et al. [25] used BAS algorithm to find the optimal value to replace the initial weight and threshold value in Elman neural network, and constructed BAS-Elman model. Compared with GA-Elman model and other models, this model has higher accuracy. Wang et al. [26] improved BAS with adaptive factors and optimized the initial threshold value of neural network with improved algorithm, which was applied to face classification. This method required short training time and high accuracy and stability. Wu et al. [27] optimized the weight between the hidden layer and the output layer of the new neural network classifier NNC by BAS algorithm, thus improving the calculation speed and prediction accuracy.

As BAS algorithm is a single individual search algorithm, the resolution ability of a single beetle to food odor concentration is not enough to enable it to find the optimal solution in the high-dimensional space, which may result in the location of beetle not updating and completely falling into the local optimal, which greatly reduce the search accuracy of BAS in the high-dimensional space.

3 Proposed Method

3.1 Deep Auto-encoder Denoising

Auto-encoder is a kind of unsupervised learning model, which is composed of an encoder and a decoder. The encoder extracts feature of the original representation and encode to a low dimension hidden representation. The decoder restores the hidden representation and generate output with the same dimension as the original representation. The training goal of the auto-encoder is to minimize reconstruction error. Since the feature dimension of the hidden layer is lower than the original feature dimension, the training process substantially enhances the feature extraction ability of the neural network to learn more dense and meaningful representations. The sparse induction characteristics of auto-encoder can be used for data denoising. Based on the modeling ability of deep network for complex data, the deep auto-encoder (DAE) can better learn different potential features in image and noise to improve the SNR and contrast of the restored image.

The structure diagram of a DAE is shown in Fig. 1. The encoder and decoder mainly use convolutional layer, maximum pooling layer, up-sampling layer, and other structures, which construct a deep network through layer stacking.

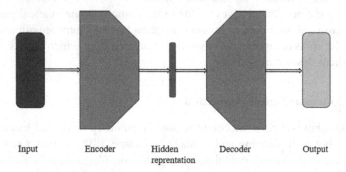

Input	Encoder	Hidden reprentation	Decoder	Output

Fig. 1. The structure diagram of a DAE.

3.2 Beetle Antennae Search Algorithm

In the process of searching for food, the beetle does not know the specific location of food. It uses the left and right antennae on its head to capture the odor concentration in the surrounding environment. If the beetle detects a higher concentration on its left antennae, it moves to the left. If not, go to the right. Through this simple principle, beetle can eventually find food efficiently. The optimization steps of the algorithm expressed by mathematical model are as follows:

The head orientation b of beetle at any position is random and normalized as:

$$b = \frac{rand(D, 1)}{|rand(D, 1)|} \tag{1}$$

where D is the spatial dimension.

(1) Left beetle antennae x_{left} and right antennae x_{right} can be expressed by the beetle head center of mass position x^m and two-antennae distance d^m as:

$$\begin{cases} x_{right} = x^m + d^m \cdot b \\ x_{left} = x^m - d^m \cdot b \end{cases} (n = 0, 1, 2, \cdots n) \tag{2}$$

(2) According to the fitness function $f(x)$, the odor concentration felt by the antennae on both sides of beetle was calculated and denoted as $f(x_{left})$ and $f(x_{right})$. The location of beetle is updated and iterated according to the following model by imitating the detection mechanism of beetle:

$$x^m = x^{m-1} + \delta^m \cdot b \cdot sign(f(x_{right}) - f(x_{left})) \tag{3}$$

(3) The updating rules of step length δ and distance d of beetle are as follows

$$\begin{cases} d^m = d_e \cdot d^{m-1} + 0.01 \\ \delta^m = \delta_e \cdot \delta^m \end{cases} \tag{4}$$

where d_e and δ_e represent the decreasing factors of d and δ respectively.

3.3 Hybrid Optimize Method

Wolpert et al. [18] point out that different optimization strategies have their own strengths. One strategy may be specially designed for the structure of a specific problem because of another strategy, and there is no universal algorithm in theory. For large-scale optimization problems, especially multi-modal, high-dimensional, constrained, and multi-objective optimization problems, only using a certain intelligent algorithm will always have its own shortcomings. Therefore, the hybrid optimization algorithm is formed by combining a variety of intelligent algorithms according to certain rules. Different algorithms make full use of their strengths and circumvent their weaknesses, which is expected to greatly improve the global and local optimization and convergence ability of the hybrid algorithm. The training of deep network itself is very time-consuming. In each iterative optimization process of the parallel hybrid algorithm, additional recalculation of the fitness function is required (depending on the parameter setting of the optimization algorithm), and the time complexity of the optimization algorithm will increase exponentially, which greatly reduce the efficiency.

In the introduction, the shortcomings of gradient descent optimization algorithm and BAS algorithm have been pointed out. This section tries to combine BAS algorithm with gradient optimization to propose a hybrid algorithm to optimize the weight parameters of deep network. There are two hybrid strategies of intelligent optimization algorithm: serial hybridize and parallel hybridize. The serial hybrid algorithm divides the number of iterations into several sections in the running process, and different optimization algorithms are used to solve the problem in turn. The parallel hybrid of algorithms divides the population into several different sub-populations, and each sub-population uses different optimization algorithm to solve. Since the hidden layer of DAE reduces the dimension of the original representation and retains as much information as possible, we propose a hybrid algorithm according to this characteristic of DAE to reduce the algorithm time consumption as much as possible. The hybrid optimize algorithm is as follows.

ALGORITHM 1: Hybrid Optimize Algorithm

---

1. Initialization

   Randomly initializes the DAE network weights W.

2. **For** n **in** $(1,2,...,N)$

   # N is the number of iterations

   # Gradient optimization stage

   Run the Gradient algorithm to optimize the DAE network weights W once.

   # BAS algorithm optimization stage

   Solidify all weights of the DAE except the hidden layer weights.

   Using BAS algorithm to optimize the hidden layer weights.

   If the convergence conditions are satisfied

   break

   end

---

4 Experiments

4.1 Training and Test

In this section, we apply our method to DAE to show the improvement of denoising performance in non-Gaussian and Gaussian noisy image. The Peak Signal to Noise Ratio (PSNR) and Structural similarity Metric (SSIM) index are calculated for evaluation. We use the noise model in [22] to synthesize the noisy images from Image data set STL [23], PolyU [24]. The images in STL and PolyU are used as training data and test data respectively. Each image is clipped to patches of size 50×50 and Gaussian and non-Gaussian noise are added to these patches multiple times to generate to generate training and testing data. Detailed network layers are shown in Table 1.

On STL, we generate training data by tow variants of noise models: (i) Gaussian noise, (ii) salt&pepper noise, (iii) heterogeneous Gaussian noise, and generate test data on PolyU and Set12 data set in the same way. The ADAM algorithm is adopted and the model is trained with 50 epochs by setting the learning rate 10–3. The number of iterations of BAS algorithm is 5 during the one epoch of the hybrid algorithm.

Figure 2 shows a denoising result on a PolyU image in AWGN noise model. Figure 3 shows a denoising result on a Set12 image in salt&pepper noise model. Table 2 lists the denoising results of three algorithms in Set12 data set and our method achieves the best PSNR/SSIM results. The original DAE methods produce artifacts and are easily over-smooth in image structures. In comparison, improved DAE performs favorably in removing noise while preserving image detail structures.

Table 1. Detailed network layers of DAE.

Encoder			Decoder		
Encoder layer	Input shape	Parameters	Decoder layer	Output shape	Parameters
Conv2d	[, 64, 96, 96]	(3, 64, 3, 1, 1)	Conv Transpose2d	[, 256, 24, 24]	(256, 128, 3,1, 1)
Batch Norm2d	[, 64, 96, 96]	64	ReLU	[, 256, 24, 24]	–
Conv2d	[, 64, 96, 96]	(64, 64, 3, 1, 1)	Batch Norm2d	[, 256, 24, 24]	128
ReLU	[, 64, 96, 96]	–	Conv Transpose2d	[, 128, 48, 48]	(128, 128, 3,2,1, 1)
MaxPool2d	[, 64, 96, 96]	(2, 2)	ReLU	[, 128, 48, 48]	–
Batch Norm2d	[, 64, 48, 48]	64	Batch Norm2d	[, 128, 48, 48]	128
Conv2d	[, 64, 48, 48]	(64, 64, 3, 1, 1)	Conv Transpose2d	[, 64, 48, 48]	(128, 64, 3,1, 1)
ReLU	[, 64, 48, 48]	–	ReLU	[, 64, 48, 48]	–
Batch Norm2d	[, 64, 48, 48]	64	Batch Norm2d	[, 64, 48, 48]	64
Conv2d	[, 64, 48, 48]	(64, 128, 3, 1, 1)	Conv Transpose2d	[, 32, 48, 48]	(64, 32, 3,1, 1)
ReLU	[, 128, 48, 48]	–	ReLU	[, 32, 48, 48]	–
Batch Norm2d	[, 128, 48, 48]	128	Batch Norm2d	[, 32, 48, 48]	32
Conv2d	[,128, 48, 48]	(128, 128, 3, 1, 1)	Conv Transpose2d	[, 32, 48, 48]	(32, 32, 3,1, 1)
ReLU	[, 128, 48, 48]	–	Conv Transpose2d	[, 16, 96, 96]	(32, 16, 3,2,1, 1)
Batch Norm2d	[, 128, 48, 48]	128	ReLU	[, 16, 96, 96]	–
Conv2d	[,256 48, 48]	(128, 256, 3, 1, 1)	Batch Norm2d	[, 16, 96, 96]	16
ReLU	[, 256, 48, 48]	–	Conv Transpose2d	[, 16, 96, 96]	(16, 3, 3,1, 1)
MaxPool2d	[, 256, 24, 24]	(2, 2)	Sigmoid	[, 3, 96, 96]	–
Batch Norm2d	[, 256, 24, 24]	256			

Table 2. The denoising results on Set12 data set under salt & pepper noise (d = 0.04).

Method	PSNR	SSIM
Origina DAE	26.36	0.8564
BM3D [28]	25.81	0.8525
Improved DAE	27.79	0.8687

Figure 4 shows a denoising result in heterogeneous Gaussian noise. In the test data set, the hybrid algorithm achieves about ~0.92 dB PSNR gain and ~0.02 SSIM gain over the original DAE. The original DAE lost some details when removing noise from complex non-Gaussian noisy photograph. While our improved method outperforms original DAE in balancing noise removal and structure preservation and achieve the highest PSNR/SSIM results.

According to the network structure, the dimension of hidden layer is 256, which has been greatly reduced compared with the input image and the whole network parameters. At the same time, the time complexity of BAS algorithm has a linear relationship with the dimension, so the increase of training time of hybrid algorithm can be ignored. It takes **2784 s** to train the original DAE on a Nvidia GeForce GTX 1660 GPU while **2805 s** using hybrid algorithm. The increasing time consumption is very little.

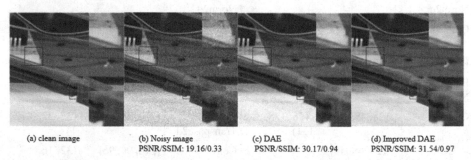

| (a) clean image | (b) Noisy image
PSNR/SSIM: 19.16/0.33 | (c) DAE
PSNR/SSIM: 30.17/0.94 | (d) Improved DAE
PSNR/SSIM: 31.54/0.97 |

Fig. 2. Denoising results of a PolyU image under Gaussian noise ($\sigma = 30$).

| (a) original image | (b) Noise image | (c) BM3D | (d) Improved DAE |

Fig. 3. Denoising results of a Set12 image under salt&pepper noise (d = 0.04).

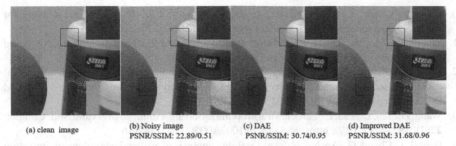

(a) clean image
(b) Noisy image
PSNR/SSIM: 22.89/0.51
(c) DAE
PSNR/SSIM: 30.74/0.95
(d) Improved DAE
PSNR/SSIM: 31.68/0.96

Fig. 4. Denoising results of PolyU image under heterogeneous Gaussian noise.

5   Conclusion

To solve the problems of slow convergence speed and easy to fall into local optimum when training high dimensional deep networks such as DAE using the gradient descent method, a new hybrid algorithm is proposed in this paper, which integrates the BAS search algorithm into the gradient descent algorithm, and the increase of algorithm time consumption is negligible. The hybrid algorithm effectively enhance the ability of the gradient descent algorithm to jump out of the local optimal. The experimental results show that the image denoising performance of DAE model trained by the hybrid algorithm has been improved effectively.

Acknowledgments. This work is sponsored by Wuchang Shouyi University Doctoral Research start up Project (No: B20200301).

References

1. Huang, Y., Bortoli, V., Zhou, F., et al.: Review of wavelet-based unsupervised texture segmentation, advantage of adaptive wavelets. IET Image Proc. **12**(9), 1626–1638 (2018)
2. Zhong, J.M., Sun, H.F.: Edge-preserving image denoising based on orthogonal wavelet transform and level sets. J. Image Graph. **6**(2), 145–151 (2018)
3. Mokari, A., Ahmadyfard, A.: Fast single image SR via dictionary learning. IET Image Proc. **11**(2), 135–144 (2017)
4. Dong, W., Li, X., Zhang, D., et al.: Sparsity-based image denoising via dictionary learning and structural clustering. In: Proceedings of IEEE Conference on Computer Vision and Pattern Recognition, Providence, RI, pp. 457–464 (2011)
5. Liu, C.S., Zhao, Z., G., Li, Q., et al.: Enhanced low-rank representation image denoising algorithm. Comput. Eng. Appl. **56**(2), 216–225 (2020)
6. Munezawa, T., Goto, T.: noise removal method for moving images using 3-D and time-domain total variation regularization decomposition. J. Image Graph. **7**(1), 18–25 (2019)
7. Jian, J., Ren, F., Ji, H.F., et al.: Generalised non-locally centralised image de-noising using sparse dictionary. IET Image Proc. **12**(7), 1072–1078 (2018)
8. Danielyan, A., Katkovnik, V., Egiazarian, K.: BM3D frameworks and variational image deblurring. IEEE Trans. Image Processing **21**(4), 1715–1728 (2018)
9. Moran, N., Schmidt, D., Zhong, Y., et al.: Nosier2Noise: learning to denoise from unpaired noisy data. arXiv: 1910.11908v1 (2019)

10. Isogawa, K., Ida, T., Shiodera, T., et al.: Deep shrinkage convolutional neural network for adaptive noise reduction. IEEE Signal Process. Lett. **25**(2), 224–228 (2017)
11. Zhang, K., Zuo, W., Zhang, L.: FFDNet: toward a fast and flexible solution for CNN based image denoising. IEEE Trans. Image Process. **27**(9), 4608–4622 (2017)
12. Chen, J., Chen, J., Chao, H., Ming, Y.: Image blind denoising with generative adversarial network based noise modeling. In: 2018 IEEE/CVF Conference on Computer Vision and Pattern Recognition (CVPR), pp. 3155–3164 (2018)
13. Wang, D., Tan, D., Liu, L.: Particle swarm optimization algorithm: an overview. Soft. Comput. **22**(2), 387–408 (2017). https://doi.org/10.1007/s00500-016-2474-6
14. Bansal, J., Chand, G., Anshul, N., et al.: Stability analysis of artificial bee colony optimization algorithm. Swarm Evol. Comput. **41**, 9–19 (2018)
15. Pan, W., T. (2012): A new fruit fly optimization algorithm: taking the financial distress model as an example. Knowl. Based Syst. **26**(2), 69–74 (2016)
16. Tran, B., Xue, B., Zhang, M.: Genetic programming for feature construction and selection in classification on high-dimensional data. Memetic Comput. **8**(1), 3–15 (2016). https://doi.org/10.1007/s12293-015-0173-y
17. Jiang, X., Li, S: BAS: beetle antennae search algorithm for optimization problems. arXiv: 1710.10724 (2017)
18. Wolpert, D.H., Macready, W.G.: No free lunch theorems for optimization. IEEE Trans. Evol. Comput. **1**(1), 67–82 (1997)
19. Vincent, P., Larochelle, H., Lajoie, I., et al.: Stacked denoising auto-encoders: Learning useful representations in a deep network with a local denoising criterion. J. Mach. Learn. Res. **11**(12), 3371–3408 (2010)
20. Zhang, K., Zuo, W., Chen, Y., et al.: Beyond a gaussian denoiser: residual learning of deep CNN for image denoising. IEEE Trans. Image Processing **26**(7), 3142–3155 (2017)
21. Guo, S., et al.: Toward convolutional blind denoising of real photographs. In: IEEE/CVF Conference on Computer Vision and Pattern Recognition, pp. 1712–1722(2019)
22. Brooks, T., et al.: Unprocessing images for learned raw denoising. In: IEEE/CVF Conference on Computer Vision and Pattern Recognition (CVPR), pp. 11036–11045 (2019)
23. Martin, D., Fowlkes, C., et al.: A database of human segmented natural images and its application to evaluating segmentation algorithms and measuring ecological statistics. In: Proceedings of IEEE Conference on Computer Vision, vol. 2, pp. 416-423 (2001)
24. Anaya, J., Barbu, A.: RENOIR - a dataset for real low-light image noise reduction. J. Vis. Commun. Image Represent. **51**, 144–154 (2014)
25. Cai, C., Qian, Q., Fu, Y.: Application of bas-elman neural network in prediction of blasting vibration velocity. Procedia Comput. Sci. **166**, 491–495 (2020)
26. Wang, F., Xie, F., Shen, S., Huang, L., et al.: A novel multiface recognition method with short training time and lightweight based on abasnet and h-softmax. IEEE Access **8**, 175370–175384 (2020)
27. Wu, Q., Ma, Z., Xu, G., Li, S., Chen, D.: A novel neural network classifier using beetle antennae search algorithm for pattern classification. IEEE Access **7**, 64686–64696 (2019)
28. Dabov, K., Foi, A., Katkovnik, V., et al.: Color image denoising via sparse 3d collaborative filtering with grouping constraint in luminance-chrominance space. In: IEEE International Conference on Image Processing, vol. 1, pp. I-313-I-316 (2007)

Diversity Technology Research in Wireless Communication Systems

Ning Ran[1], Lina Yuan[1(✉)], and Anran Zhou[2]

[1] College of Data Science, Tongren University, Tongren 554300, China
dsjyln@gztrc.edu.cn
[2] College of Computer Science, Chengdu University, Chengdu 6100059, China

Abstract. With the continuous development of science and technology and wireless communication, wireless technology is in a stage of rapid development, and various fields have been widely used, for instance, 5G, Internet of Things, mobile edge computing, and wireless powered communication. Therefore, wireless communication is more and more useful, the quality of transmission has been continuously improved, however, the fading problem in the channel has always been the key factor limiting its performance. Hence, diversity technology as an effective anti-fading means, its own research needs to be more in-depth. In this paper, the basic principle and classification of diversity technology and three kinds of diversity combining technology are investigated, and four scenarios of maximum-ratio combining schemes and bit error rate performance using Binary Phase Shift Keying (BPSK), Quadrature Phase Shift Keying (QPSK), 8QPSK and 16 Quadrature Amplitude Modulation (16QAM) are simulated in Rayleigh fading channel, respectively.

Keywords: Wireless communication · Diversity technology · Maximum-Ratio Combining · Bit error rate · Rayleigh fading

1 Introduction

With the wide application of wireless communication technology in various fields, how to achieve reliable signal transmission in the complex and changeable environment has become a hot issue in wireless communication [1]. In the process of wireless signal transmission, due to the influence of external environment, such as atmospheric refraction, building diffraction, sea reflection and movement, the signal will decline. However, Diversity technology has unique advantages in resisting fading, especially in large-capacity and long-distance communication. Generally speaking, if there is a fading problem in wireless communication, part of the special sequence code received by the receiver will be lost, and the location and recovery of the signal cannot be accurately realized. Diversity technology [1] is equivalent to the replication and transmission of information. It makes full use of the unrelated characteristics in time, space and frequency domains, and can reasonably integrate signals in the case of fast fading to ensure reliable communication, so as to achieve accurate timing and synchronization of the

F. Neri et al. (Eds.): CCCE 2022, CCIS 1630, pp. 85–93, 2022.
https://doi.org/10.1007/978-3-031-17422-3_8

system. After diversity technology is put forward, some key technologies also come into being, among which the main technology includes diversity combining reception.

Diversity technology can be applied to 5G [2], wireless powered communication [3–5], mobile edge computing [6] and Internet of Things [7], whose application requires comprehensive consideration and analysis of its anti-fading effect, implementation mode and various channel models of the system. In Fundamentals of Communication [1], D. Tse and P. Viswanath divided diversity techniques into frequency, time, and space diversity. In some specific channel environments, many international scholars have obtained the performance comparison of various diversity technologies in different channel environments by comparing the performance of various diversity technologies. In addition, many scholars combined digital communication with digital signal to improve the quality of signal before and after combing, and finally achieved the overall effect of improving diversity. B. Hart proposed a method to achieve diversity reception through LLR estimation [8]. In 1999, A. M. Sayeed and B. Aazhang [9] proposed to obtain mutually independent diversity branches through the Doppler effect of channels, thus providing a new direction for the development of diversity technology.

2 Diversity Technology

2.1 The Basic Principle

Diversity technology is a common technique in scattering communication. Diversity technology can effectively improve the gain of the system and enhance the reliability of communication. Diversity refers to the transmission of the same information through two or more ways to reduce the impact of fading.

Diversity technology mainly includes two aspects: one is "dividing", that is to send multiple copies of the same data at the transmitter, so that the receiver can get different quality signals carrying the same information; On the other hand, it is "combing". Each signal reaches the receiver through different fading, and the gain and anti-fading ability of the system can be improved by combining the received signals. Therefore, one of the most important conditions of diversity technology is to ensure the irrelevance of each signal.

2.2 The Classification

Diversity technology is used to mitigate the decline of error performance caused by unstable fading of wireless channels (such as multipath fading) [10, 11]. Diversity in data transmission is mainly based on the following idea: the probability of multiple statistically independent fading channels simultaneously in deep fading is very low. Diversity technology can be divided into many kinds according to different types, but they are basically considered from the three aspects of time domain, space domain and frequency domain.

Time diversity [11] means that the same signal carrying the same data will be retransmitted on the time axis for many times, but it needs to ensure that the time interval of each re-transmission is greater than the coherent time of the channel, that is, the time interval of re-transmission should be satisfied

$$\Delta t > \frac{1}{2f_m} = \frac{1}{2(v/\lambda)}, \tag{1}$$

Here f_m is the operating frequency, and λ is the operating wavelength. Time diversity mainly uses the irrelevance of signal fading in time to achieve time selective fading. Time diversity is relatively easy to achieve, without considering much cost and occupation of frequency band resources. However, because time diversity is to resend data within a certain time interval, more storage space is needed for data storage, and the receiving delay is large.

In the space domain, spatial diversity is mainly used [11], which is also known as antenna diversity. This diversity mode is mainly realized by using two pairs or even more antennas, and generally requires that the distance between adjacent antennas $d > \lambda/2$ (λ is operating wavelength). To ensure that is not related to the fading of received signal characteristics, that is, when one pair of antenna reception when the signal-to-noise ratio (SNR) is very low, at the same time, other antenna reception SNR can be high, so as to realize the purpose of sharing "risk". The advantage of spatial diversity is that the diversity gain is high, but the disadvantage is that it requires multiple antennas and the cost is high. Polarization diversity and angular diversity belong to two special cases of spatial diversity.

Frequency diversity [8, 11] refers to the fact that signals in different frequency bands are mathematically statistically irrelevant after passing through fading channels, that is to say, selective fading of frequencies can be achieved by utilizing differences in fading statistical characteristics in different frequency bands with a certain interval. In general, the frequency interval between the two frequencies should be greater than the coherence interval of the frequencies, i.e., If the first chosen antenna is given, the second antenna can be chosen to maximize the channel capacity [9]:

$$\Delta f > B_c = \frac{1}{\Delta \tau_m}, \tag{2}$$

Here Δf is carrier frequency interval, B_c is correlation bandwidth, and $\Delta \tau_m$ is the maximum multipath delay inequality. The biggest difference between frequency diversity and spatial diversity is that the former does not need multiple antennas to receive signals, but needs to occupy more frequency band resources. Therefore, in most cases, frequency diversity is also called in-band diversity.

3 Diversity Combining Technology

Diversity technology is a kind of research on how to use the multipath signal in transmission to enhance the reliability of transmission. In addition, diversity technology is also a study of signal transmission and reception technology. Diversity of "dividing" and "combing" itself is a kind of opposite. At the transmitting end, it is necessary to disperse transmission in various ways, and at the receiving end, it is necessary to superimpose N independent signals through certain combining methods and output them, so as to maximize the gain. The method of diversity merging is generally by adding weights [11]. Diversity Combining technology can be summarized as: Combine multiple received signals in a diversity receiving device into a single improved signal. Diversity Combining technology can be divided into three types, respectively: Selection Combining (SC), Equal Gain Combining (EGC) and Maximum-Ratio Combining (MRC). These three technologies are briefly described below.

Consider a single input multiple output system model, the received vector signal is

$$y(n) = hx(n) + z(n), \tag{3}$$

here, $y(n) = [y_1(n), \cdots, y_M(n)]^T$, channel gain matrix is $h = [h_1, \cdots, h_M]^T$, $h_i = \sqrt{\beta_i} e^{j\theta_i}$, $i = 1, \cdots, M$ ($\beta_i = |h_i|^2$ represents amplitude square or power, and $\theta_i = \angle h_i$ is phase). Let P_t represent transmitting power and $E[|x(n)|^2] = P_t$. $z(n) = [z_1(n), \cdots, z_M(n)]^T$, $z_i(n) \sim CN(0, \sigma^2)$ and $z_i(n)$ is mutually independent and identically distributed complex Gaussian noise. Introduce a new variable, γ_i denotes the i-th branch on the SNR, defined as $\gamma_i = \beta_i P_t / \sigma^2$. Then it can be known that the combined output signal is

$$y_\Sigma(n) = w^T y(n) = \left(\sum_{i=1}^M w_i h_i \right) x(n) + \sum_{i=1}^M w_i z_i(n). \tag{4}$$

And the combined output SNR is

$$\gamma_\Sigma = \frac{E\left[\left| \left(\sum_{i=1}^M w_i h_i \right) x(n) \right|^2 \right]}{E\left[\left| \sum_{i=1}^M w_i z_i(n) \right|^2 \right]} = \frac{\left(\sum_{i=1}^M a_i \sqrt{\beta_i} \right)^2 P_t}{\left(\sum_{i=1}^M a_i^2 \right) \sigma^2}, \tag{5}$$

here $a_i = w_i$ denotes weight.

3.1 Selection Combining (SC)

SC selects the signal with the largest SNR branch as the output of the combiner, and the combining weight is

$$a_i^{SC} = \begin{cases} 1, & \text{if } i = \arg\max_{j} \gamma_j \\ 0, & \text{otherwise} \end{cases} \quad i = 1, \cdots, M. \tag{6}$$

Then the combined output SNR is

$$\gamma_{SC} = \max(\gamma_1, \cdots, \gamma_M) = \max\left(\frac{\beta_i P_t}{\sigma^2}\right). \tag{7}$$

3.2 Equal Gain Combining (EGC)

EGC is the consistent summation of all signals received by the receiver. For SC, only one signal on one branch is used at a given time; For EGC, signals in all branches are combined with equal weight, and the combined weight is

$$a_i^{EGC} = 1, \ i = 1, \cdots, M. \tag{8}$$

According to $\gamma_{\Sigma} = \dfrac{\left(\sum\limits_{i=1}^{M} a_i \sqrt{\beta_i}\right)^2 P_t}{\left(\sum\limits_{i=1}^{M} a_i^2\right)\sigma^2}$, the combined output SNR is

$$\gamma_{EGC} = \frac{\left(\sum\limits_{i=1}^{M} \sqrt{\beta_i}\right)^2 P_t}{M\sigma^2}. \tag{9}$$

3.3 Maximum-Ratio Combining (MRC)

MRC is often used in large-scale proportional array systems, where SNR received is weighted and then summed. The SNR generated is $\sum\limits_{k=1}^{N} SNR_k$, here SNR_k is the SNR of the i-th signal received. EGC applies equal weights to signals in all combing branches, and for MRC, the combined weight is chosen to maximize the output SNR, and the combined weight is

$$a_i^{MRC} = \sqrt{\beta_i}, \ i = 1, \cdots, M. \tag{10}$$

According to $\gamma_{\Sigma} = \dfrac{\left(\sum\limits_{i=1}^{M} a_i \sqrt{\beta_i}\right)^2 P_t}{\left(\sum\limits_{i=1}^{M} a_i^2\right) \sigma^2}$, the combined output SNR is

$$\gamma_{MRC} = \frac{\left(\sum\limits_{i=1}^{M} \sqrt{\beta_i}\sqrt{\beta_i}\right)^2 P_t}{\sum\limits_{i=1}^{M} \left(\sqrt{\beta_i}\right)^2 \sigma^2} = \sum_{i=1}^{M} \frac{\beta_i P_t}{\sigma^2}. \tag{11}$$

Note $\gamma_{MRC} = \sum\limits_{i=1}^{M} \gamma_i$, it has a very nice property that the output SNR is the sum of the SNR of all the independent branches. According to the Cauchy-Schwarz inequality:

$$\left(\sum_{i=1}^{M} a_i \sqrt{\beta_i}\right)^2 \leq \sum_{i=1}^{M} a_i^2 \sum_{i=1}^{M} \left(\sqrt{\beta_i}\right)^2. \tag{12}$$

If and only if $a_i / \sqrt{\beta_i} = c$, $\forall i$ and c is any normal number, the equation above is equal.

In telecommunication, the MRC method in the diversity combination mentioned above is mainly applied in three aspects: Firstly, the signals of each channel are overlapped together; Secondly, the gain of each channel is proportional to the root mean square signal and inversely proportional to the mean square noise; Finally, different proportionality constants are used in each channel.

4 Numerical Simulation Results

This section compares the bit error rate (BER) of four sceneries with varying the SNR (dB) from 0 dB to 20 dB for four schemes in Rayleigh fading. In the following simulation, frame length is 100 s, and the number of packets is 5000 [11]. Figure 1 demonstrates the BER performance of MRC for four schemes, i.e., SISO ($N_{Tx} = 1$ and $N_{Rx} = 1$), MRC ($N_{Tx} = 1$ and $N_{Rx} = 2$), MRC ($N_{Tx} = 1$ and $N_{Rx} = 4$) and MRC ($N_{Tx} = 1$ and $N_{Rx} = 8$), meanwhile adopts Binary Phase Shift Keying (BPSK) Modem. It is observed from Fig. 2 that all schemes' BER decreases with the increase of SNR. However, the BER performance of MRC ($N_{Tx} = 1$ and $N_{Rx} = 8$) is the least, and SISO ($N_{Tx} = 1$ and $N_{Rx} = 1$) is the worst. Similarly, the same simulation is done with Quadrature Phase Shift Keying (QPSK), 8QPSK and 16 Quadrature Amplitude Modulation (16QAM) modulators, and the same results are obtained as shown in Fig. 2, Fig. 3, and Fig. 4, respectively.

For Additive White Gaussian Noise (AWGN) channels, when SNR is large, the slope of BER curve tends to infinity, that is to say, when SNR increases, the channel presents a kind of BER performance of falling water. For Rayleigh fading channels, the corresponding slope of the curve is linear in log-log coordinates. The main purpose of antenna diversity technology is to transform the unstable wireless time-varying fading channel into a stable channel without significant instantaneous fading like AWGN channel.

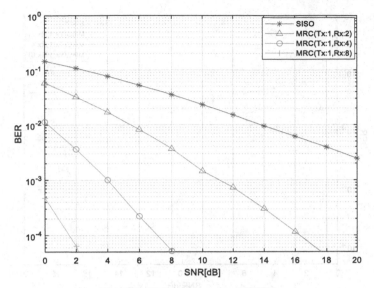

Fig. 1. MRC performance of BPSK.

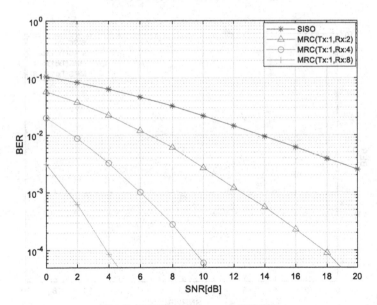

Fig. 2. MRC performance of QPSK.

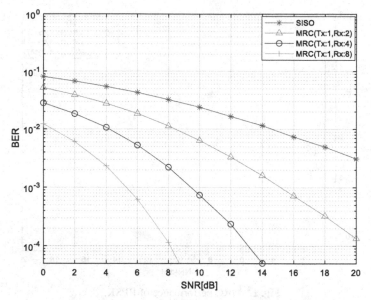

Fig. 3. MRC performance of 8QPSK.

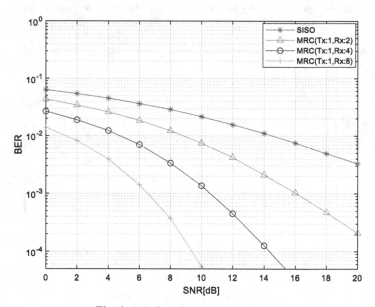

Fig. 4. MRC performance of 16QAM.

5 Summary

The research of diversity technology is a heavy task and a long way to go. So far, many researches are only established in the ideal laboratory environment, lacking the support of engineering practice. At the same time, a single diversity method may not be able to achieve the characteristics of anti-fading, and the complexity is relatively high, so in a specific use environment, the interaction of two or more diversity schemes will be the focus and difficulty of diversity technology research, and also the main direction of our future research.

Acknowledgment. Thanks are due to the referees for helpful comments. This research was supported by Doctoral research project of Tongren University (trxyDH2003), Doctoral talent project of Tongren Science and Technology Bureau (No. [2020]124), Basic Research Program of Guizhou Province-ZK[2021] General 299, and Science and Technology Planning Project of Guizhou Province-[2020]1Y260.

References

1. Tse, D., Viswanath, P.: Fundamentals of wireless communication. IEEE Trans. Inf. Theory **55**(2), 919–920 (2009)
2. Romeo, F., Campolo, C., Berthet, A.O., et al.: Improving the DENM reliability over 5G-V2X Sidelink through repetitions and diversity combining. In: 2021 IEEE 4th 5G World Forum (5GWF), Montreal, QC, Canada, pp. 352-357 (2021)
3. Bi, S., Ho, C.K., Zhang, R.: Wireless powered communication: opportunities and challenges. IEEE Commun. Mag. **53**(4), 117–124 (2015)
4. Bi, S., Zeng, Y., Zhang, R.: wireless powered communication networks: an overview. IEEE Wirel. Commun. Mag. **23**(2), 1536–1683 (2016)
5. Yuan, L., Chen, H., Gong, J.: Interactive communication with clustering collaboration for wireless powered communication networks. Int. J. Distrib. Sens. Netw. **18**(2), 1–10 (2022)
6. Bi, S., Zhang, Y.J.: Computation rate maximization for wireless powered mobile-edge computing with binary computation offloading. IEEE Trans. Wireless Commun. **17**(6), 4177–4190 (2018)
7. Wen, M., Lin, S., Kim, K.J., et al.: Cyclic delay diversity with index modulation for green internet of things. IEEE Trans. Green Commun. Netw. **5**(2), 600–610 (2021)
8. Hart, B., Taylor, D.P.: Extended MLSE diversity receiver for the time and frequency selective channel. IEEE Trans. Commun. **45**(3), 322–333 (1997)
9. Sayeed, A.M., Aazhang, B.: Joint multipath-doppler diversity in mobile wireless communications. IEEE Trans. Commun. **47**(1), 123–132 (1999)
10. Traveset, J.V., Caire, G., Biglieri, E., et al.: Impact of diversity reception on fading channels with coded modulation-part i: coerent detection. IEEE Trans. Commun. **45**(5), 563–572 (1997)
11. Yong, S.C., Jaekwon, K., Won, Y.Y., et al.: MIMO-OFDM Wireless Communications with MATLAB, Publishing House of Electronics Industry, pp. 240–245 (2013)

Study on Solving Algorithm of T Matrix Eigenvalues Based on C Programming Language

Weida Qin, Ricai Luo$^{(\boxtimes)}$, and Yinhu Wei

Hechi University, Hechi, China
hcxylor@126.com

Abstract. The problem of solving matrix eigenvalues is a hot research topic, and dichotomy is an important algorithm to solve it. On the basis of the research on matrix eigenvalue algorithm by experts and scholars, the T matrix is studied and found to have eight pairs of eigenvalues of adjacent two sequential principal square submatrices that are not strictly interlaced, that is, the matrix does not meet the key properties of using dichotomy to solve eigenvalues. Then, the feasible conditions and feasibility of dichotomy algorithm for solving the eigenvalue of T matrix are studied by analyzing the principle of sign change number of characteristic polynomial sequence of sequential principal square submatrices of T matrix. Finally, the accuracy of the algorithm is verified by numerical simulation using C programming language.

Keywords: T matrix · Dichotomy · Eigenvalue · C programming language · Algorithm research

1 Introduction

Eigenvalues have an important application in the field of vibration problem [1–3]. There are many algorithms to solve matrix eigenvalue such as clustering algorithm, etc. [4–6], and dichotomy is an important one [7].

Let the eigenvalues of k-order sequential principal square submatrices R_k of n-order matrix R be $\lambda_1, \lambda_2, \ldots, \lambda_k$, and the eigenvalues of the $k + 1$-order sequential principal square submatrices R_{k+1} be $\mu_1, \mu_2, \ldots, \mu_{k+1}$. In the reference [7], it is pointed out that the key property of using dichotomy to solve the eigenvalues of a matrix is that the eigenvalues of any two adjacent sequential principal square submatrices of the matrix are strictly interlaced, i.e. for $k = 1, 2, \ldots, n - 1$, inequality (1) holds

$$\mu_1 < \lambda_1 < \mu_2 < \lambda_2 < \mu_3 \ldots < \mu_k < \lambda_k < \mu_{k+1}. \tag{1}$$

If there is a positive integer i such that $q_i e_i > 0 (i = 1, \cdots, n - 9)$ and $q_{n-8} e_{n-8} = q_{n-7} e_{n-7} = \cdots = q_{n-1} e_{n-1} = 0$, the n-order matrix R is called a T matrix.

F. Neri et al. (Eds.): CCCE 2022, CCIS 1630, pp. 94–103, 2022.
https://doi.org/10.1007/978-3-031-17422-3_9

Supposing the j-order sequential principal square submatrix R of T matrix is R_j, then $R_1 = b_1, R_n = R$. If the characteristic polynomial of R_j is $p_j(x)$, then

$$p_j(x) = \det(xI - R_j)$$

$$= \begin{vmatrix} x - w_1 & -e_1 & & 0 \\ -q_1 & x - w_2 & \ddots & \\ & \ddots & \ddots & -e_{j-1} \\ 0 & & -q_{j-1} & x - w_j \end{vmatrix}$$

$$= (x - w_j)p_{j-1}(x) + e_{j-1} \begin{vmatrix} x - w_1 & -e_1 & & & \\ -q_1 & x - w_2 & -e_2 & & \\ & \ddots & \ddots & \ddots & \\ 0 & & -q_{j-3} & x - w_{j-2} & -e_{j-2} \\ 0 & & 0 & 0 & -q_{j-1} \end{vmatrix} \tag{2}$$

$$= (x - w_j)p_{j-1}(x) - q_{j-1}e_{j-1} \begin{vmatrix} x - w_1 & -e_1 & & 0 \\ -q_1 & x - w_2 & \ddots & \\ & \ddots & \ddots & -e_{j-3} \\ 0 & & -q_{j-3} & x - w_{j-2} \end{vmatrix}$$

$$= (x - w_j)p_{j-1}(x) - q_{j-1}e_{j-1}p_{j-2}(x).$$

Let $p_0(x) = 1$, and according to Eq. (2) $j = 2, 3, ..., n$, there will be

$$p_j(x) = (x - w_j)p_{j-1}(x) - q_{j-1}e_{j-1}p_{j-2}(x). \tag{3}$$

According to the definition of T matrix,

$$\because q_{n-8}e_{n-8} = q_{n-7}e_{n-7} = ... = q_{n-1}e_{n-1} = 0,$$

Bases on Eq. (3), there will be

$$p_{n-7}(x) = (x - w_{n-7})p_{n-8}(x) \tag{4}$$

$$p_{n-6}(x) = (x - w_{n-6})p_{n-7}(x) \tag{5}$$

$$p_{n-5}(x) = (x - w_{n-5})p_{n-6}(x) \tag{6}$$

$$p_{n-4}(x) = (x - w_{n-4})p_{n-5}(x) \tag{7}$$

$$p_{n-3}(x) = (x - w_{n-3})p_{n-4}(x) \tag{8}$$

$$p_{n-2}(x) = (x - w_{n-2})p_{n-3}(x) \tag{9}$$

$$p_{n-1}(x) = (x - w_{n-1})p_{n-2}(x) \tag{10}$$

$$p_n(x) = (x - w_n)p_{n-1}(x) \tag{11}$$

Equations (4–11) show that a T matrix has eight pairs of characteristic polynomials $p_{l-1}(x)$ and $p_l(x)(l = n - 7, n - 6, \ldots, n)$ that have a common root for two adjacent sequential primary square submatrices, that is, the roots of $p_{l-1}(x)$ and $p_l(x)(l = n - 7, n-6, \ldots, n)$ are not strictly interlaced. Thus, T matrix does not meet the key properties of solving matrix eigenvalues by dichotomy.

In reference [8], it is pointed out that dichotomy will be the most suitable algorithm for calculating partial eigenvalues of specified intervals. In view of the important position of dichotomy in algorithmic problem solving, the necessary conditions and feasibility of it for solving eigenvalues of T matrix was studied in this paper. Finally, the algorithm was verified by a numerical experiment using C programming language.

2 Definition and Lemma

Definition 1 [9]: A sequence of real numbers

$$b_0, b_1, \ldots, b_{m-2}, b_{m-1}, b_m,$$

and $b_0 \neq 0$, $b_m \neq 0$, if b_{i-1} and b_i are opposite in sign counting from left to right, it means that there is a sign change number; if $b_i = 0$, b_{i-1} and b_{i+1} are opposite in sign, there is also a sign change number, which is the sum of the sign change numbers in the sequence.

According to the definition, count the sequence of real numbers $-12, -33, 3, 6, -5, 2, -9, -8, 12$.

From left to right, there are 5 sign change numbers, namely, -33 and 3, 6 and -5, -5 and 2, 2 and -9, -8 and 12.

Count the sequence of real numbers $-3, -21, 0, 37, -5$ from left to right, there are 2 sign change numbers, namely, -21 and 37, 37 and -5.

Definition 2 [9]: Let $f_0(x), f_1(x), \ldots, f_m(x)$ be a polynomial with real coefficients, and x_0 be a real, in the sequence of real numbers

$$f_0(x_0), f_1(x_0), \ldots, f_m(x_0)$$

there is $f_0(x_0) \neq 0, f_m(x_0) \neq 0$, then the sign change numbers of this sequence is called the sign change numbers of this polynomial sequence at x_0.

Definition 3 [10]: If the real number x_0 is a root of $f_0(x)$, and there is a positive number ε such that $f_0(x)f_1(x) > 0$ or $f_0(x)f_1(x) < 0$ constantly for any $x \in (x_0 - \varepsilon, x_0)$, then x_0 is called generalized zero of the third kind of $f_0(x)$ about $f_1(x)$.

Definition 3 [9]: Let q_i, w_i, e_i be reals

$$H = \begin{bmatrix} w_1 & e_1 & & 0 \\ q_1 & w_2 & \ddots & \\ & \ddots & \ddots & e_{n-1} \\ 0 & & q_{n-1} & w_n \end{bmatrix}.$$

If for all $i = 1, \ldots, n - 1, q_i e_i > 0$, H is a Jacobian matrix.

Supposing the j-order sequential principal square submatrix of H is H_j, and its characteristic polynomial is $p_j(x)$. Let $p_0(x) = 1$, and according to Eq. (2) $j = 2, 3, \ldots, n$, there will be

$$p_j(x) = (x - w_j)p_{j-1}(x) - q_{j-1}e_{j-1}p_{j-2}(x)(q_{j-1}e_{j-1} > 0) \tag{12}$$

The sequence $p_0(x), p_1(x), \ldots, p_{n-1}(x), p_n(x)$ satisfying Eq. (12) is called the characteristic polynomial sequence for the sequential primary square submatrix of the n-order Jacobian matrix.

Lemma 1 [10]: The dichotomy for solving the eigenvalue of Jacobi matrix is: Let $p_0(x) = 1$, and V(x) be the sign change number of characteristic polynomial sequence $p_0(x), p_1(x), \ldots, p_n(x)$ for the sequential primary square submatrix of the n-order Jacobian matrix, with given a, b, if $p_n(a) \neq 0, p_n(b) \neq 0$, then $V(a) - V(b)$ is the number of roots of $p_n(x)$ in the interval $[a, b]$.

3   Theorem of T Matrix Dichotomy

Supposing the j-order sequential principal square submatrix of T matrix R is R_j, and its characteristic polynomial is $p_j(x)$. Let $p_0(x) = 1$, and according to Eq. (2) $j = 2, 3, \ldots, n$, there will be

$$p_j(x) = (x - w_j)p_{j-1}(x) - q_{j-1}e_{j-1}p_{j-2}(x). \tag{13}$$

According to the definition of T matrix, $\because q_i e_i > 0(i = 1, \cdots, n - 9)$, then

$$p_0(x) = 1,$$

$$p_1(x) = x - w_1,$$

$$p_2(x) = (x - w_2)p_1(x) - q_1 e_1 p_0(x),$$

$$p_3(x) = (x - w_3)p_2(x) - q_2 e_2 p_1(x)$$

$$\cdots,$$

$$p_{n-8}(x) = (x - w_{n-8})p_{n-9}(x) - q_{n-9}e_{n-9}p_{n-10}(x),$$

According to Eq. (12), $p_0(x), p_1(x), \ldots, p_{n-8}(x)$ is a characteristic polynomial sequence for the sequential primary square submatrix of the n-8-order Jacobian matrix.

According to Eqs. (4–11)

$$p_n(x) = (x - w_{n-7})(x - w_{n-6}) \ldots (x - w_{n-1})(x - w_n)p_{n-8}(x). \tag{14}$$

Supposing $S(x)$ is the sign change number of the sequence $p_0(x), p_1(x), \ldots, p_n(x)$, $V(x)$ is the sign change number of the sequence $p_0(x), p_1(x), \ldots, p_{n-8}(x)$, and $U(x)$ is the sign change number of the sequence $p_{n-8}(x), \ldots, p_{n-1}(x), p_n(x)$, then $S(x) = V(x) + U(x)$.

Theorem 1. If $W_{n-7}, W_{n-6}, \ldots W_n \notin [a, b]$, let $p_0(x) = 1$, and $S(x)$ be the sign change number of characteristic polynomial sequence $p_0(x), p_1(x), \cdots, p_n(x)$ for the sequential primary square submatrix of the n-order T matrix, with given a, b, if $p_n(a) \neq 0, p_n(b) \neq 0$, then $S(a) - S(b)$ is the number of roots of $p_n(x)$ in the interval $[a, b]$.

$$\text{Prove } S(a) - S(b) = V(a) + U(a) - (V(b) + U(b)) = V(a) - V(b)) + (U(a) - U(b)). \tag{15}$$

According to Eq. (14) and $p_n(a) \neq 0, p_n(b) \neq 0$, then $p_{n-8}(a) \neq 0, p_{n-8}(b) \neq 0$.

Supposing d is the number of roots of $p_{n-8}(x)$ in the interval $[a, b]$, and because $p_0(x), p_1(x), \ldots p_{n-8}(x)$ is the characteristic polynomial sequence for the sequential primary square submatrix of the n-8-order Jacobian matrix, according to lemma 1, $V(a) - V(b) = d$. $\because w_{n-7}, w_{n-6}, \ldots, w_n \notin [a, b]$, the number of roots of $p_n(x)$ and $p_{n-8}(x)$ are equal in $[a, b]$ according to Eq. (15). According to Eq. (15) and $V(a) - V(b) = d$, so the theorem will be true only when $U(a) - U(b) = 0$ is proved to be true.

Put all the roots of $p_{n-8}(x), \ldots, p_{n-1}(x), p_n(x)$ in $[a, b]$ from small to large

$$h_1, h_2, \ldots, h_{m-1}, h_m$$

Let $h_0 = a, h_{m+1} = b$, then there are $m + 1$ disjoint short intervals in $[a, b]$

$$(h_0, h_1), (h_1, h_2), \cdots, (h_m, h_{m+1}).$$

Every polynomial $p_j(x)$ $(j = n - 8, \ldots, n - 1, n)$ in the sequence has the same sign in every short interval (h_i, h_{i+1}), i.e., $x \in (h_i, h_{i+1})$, $U(x)$ is a constant. Take any point x_i in each short interval (h_i, h_{i+1}), and $U(x_i)$ represents the sign change number of the sequence $p_{n-8}(x), \ldots, p_{n-1}(x), p_n(x)$ in (h_i, h_{i+1}), and then there will be a sign change number sequence

$$U(a), U(x_0), U(x_1), \ldots, U(x_n), U(b).$$

Mark $\Delta_i = U(x_i) - U(x_{i+1})$, then

$$U(a) - U(b) = U(a) - U(x_0) + U(x_0) - U(x_1) + \cdots + U(x_{m-1}) - U(x_m) + U(x_m) - U(b)$$

$$= U(a) - U(x_0) + \sum_{i=0}^{m-1} \Delta_i + U(x_{m+1}) - U(b).$$

The values of Δ_i and $U(a) - U(x_0), U(x_m) - U(b)$ will be analyzed hereunder in 3.1, 3.2 and 3.3.

(3.1) Firstly, $\Delta_i = U(x_i) - U(x_{i+1})(i = 0, \ldots, m - 1)$ is analyzed, i.e., the change of $U(x)$ when x is changed from (h_i, h_{i+1}) to (h_{i+1}, h_{i+2}). According to Eqs. (4–11), every h_{i+1} is the root of $p_n(x)$. And according to Eq. (14) and $w_{n-7}, w_{n-6}, \ldots, w_n \notin [a, b]$, we can get $p_{n-8}(h_{i+1}) = 0$.

$\because p_{n-8}(h_{i+1}) = 0$, so h_{i+1} is also the root of $p_{n-7}(x), \ldots, p_{n-1}(x), p_n(x)$ according to Eqs. (4–11).

When $x \in (h_i, h_{i+1}) \cup (h_{i+1}, h_{i+2})$, $\because w_{n-7}, w_{n-6}, \ldots, w_n \notin [a, b]$, then we can get $h_{i+1} \neq w_{n-7}, \ldots, w_{n-1}, w_n$. According to $p_l(x)p_{l+1}(x) = (x - w_{l+1})p_l^2(x)(l = n - 8, \ldots, n - 2, n - 1)$, the product of $p_l(x)$ and $p_{l+1}(x)$ is greater or less than 0 constantly. That is, when $x \in (h_i, h_{i+1})$, if the product of $p_l(x)$ and $p_{l+1}(x)$ is greater than 0, that of

$p_l(x)$ and $p_{l+1}(x)$ will also be greater than 0 when $x \in (h_{i+1}, h_{i+2})$, and vice versa, thus h_{i+1} is the generalized root of the third kind of $p_{l+1}(x)$ about $p_l(x)$. As a result, the sign change number of $p_l(x_i), p_{l+1}(x_i)$ are the same with that in $p_l(x_{i+1}), p_{l+1}(X_i)$, and the sign change number of $p_{n-8}(x_i), \ldots, p_{n-1}(x_i), p_n(x_i)p_l(x_i), p_{l+1}(x_i)$ are the same with that in $p_{n-8}(x_{i+1}), \ldots, p_{n-1}(x_{i+1}), p_n(x_{i+1})$, i.e., $U(x_i) = U(x_{i+1})$,

$$U(x_i) - U(x_{i+1}) = 0.$$

Then $\sum_{i=0}^{m-1} \Delta_i = 0$.

(3.2) Since a is not a root of $p_n(x)$, it is not the root of $p_{n-8}(x), \ldots, p_{n-2}(x), p_{n-1}(x)$ either according to Eqs. (4–11).

$$p_l(x)p_{l+1}(x) = (x - w_{l+1})p_l^2(x)(1 = n - 8, \ldots, n - 2, n - 1),$$

If a is not a root of them, the sign change numbers of $p_l(x), p_{l+1}(x)$ at $a = h_0$ will be the same with those in (h_0, h_1), i.e., sign change numbers of $p_{n-8}(x), \ldots, p_{n-1}(x), p_n(x)$ at a are the same with those in (h_0, h_1), then we can get $U(a) - U(x_0) = 0$.

(3.3) Since b is not a root of $p_n(x)$, it is not the root of $p_{n-8}(x), \ldots, p_{n-2}(x), p_{n-1}(x)$ either according to Eqs. (4–11).

$$p_l(x)p_{l+1}(x) = (x - w_{l+1})p_l^2(x)(l = n - 8, \ldots, n - 2, n - 1),$$

If b is not a root of them, the sign change numbers of $p_l(x), p_{l+1}(x)$ at $b = h_{m+1}$ will be the same with those in (h_m, h_{m+1}), i.e., sign change numbers of $p_{n-8}(x), \ldots, p_{n-1}(x), p_n(x)$ at b are the same with those in (h_m, h_{m+1}), then we can get $U(x_m) - U(b) = 0$.

$$\because U(a) - U(b) = U(a) - U(x_0) + \sum_{i=0}^{m-1} \Delta_i + U(x_n) - U(b),$$

Based on (3.1), (3.2) and (3.3), we can get

$$U(a) - U(b) = 0.$$

4 Numerical Experiment and Programming

Firstly, the expressions and eigenvalues of T matrix examples are obtained, then Java is used to design the program, and finally the running results of the program are analyzed.

4.1 Using Python Command to Find the Eigenvalue of T Matrix

See Fig. 1 for an example of using Python command to find the eigenvalue of Q matrix.

```
]:  import numpy as np
    A = np.array([[10, 1, 0, 0, 0, 0, 0, 0, 0, 0], [1, 9, 0, 0, 0, 0, 0, 0, 0, 0],
                  [0, 1, 8, 1, 0, 0, 0, 0, 0, 0], [0, 0, 0, 7, 0, 0, 0, 0, 0, 0],
                  [0, 0, 0, 6, 6, 0, 0, 0, 0, 0], [0, 0, 0, 0, 5, 5, 0, 0, 0, 0],
                  [0, 0, 0, 0, 0, 4, 4, 0, 0, 0], [0, 0, 0, 0, 0, 0, 3, 3, 0, 0],
                  [0, 0, 0, 0, 0, 0, 0, 2, 2, 0], [0, 0, 0, 0, 0, 0, 0, 0, 1, 1]])
    print('print A: \n{}'.format(A))
    a, b = np.linalg.eig(A)
    print('print eigenvalue a: \n{}'.format(a))
```

```
print A:
[[10  1  0  0  0  0  0  0  0  0]
 [ 1  9  0  0  0  0  0  0  0  0]
 [ 0  1  8  1  0  0  0  0  0  0]
 [ 0  0  0  7  0  0  0  0  0  0]
 [ 0  0  0  6  6  0  0  0  0  0]
 [ 0  0  0  0  5  5  0  0  0  0]
 [ 0  0  0  0  0  4  4  0  0  0]
 [ 0  0  0  0  0  0  3  3  0  0]
 [ 0  0  0  0  0  0  0  2  2  0]
 [ 0  0  0  0  0  0  0  0  1  1]]
print eigenvalue a:
[ 8.          8.38196601 10.61803399  1.          2.          3.
  4.          5.          6.          7.         ]
```

Fig. 1. An example of using Python command to find the eigenvalue of T matrix.

4.2 Using C Language to Design T Matrix Dichotomy Program

According to Theorem 1, the program of T matrix dichotomy is designed by using C language. The program can calculate the values of variable sign numbers s (a), s (b) and $p_n(a), p_n(b)$ by setting the interval [a, b] as [9], [11].

```
#include <stdio.h>
#include <stdlib.h>
int main()
{
    float e[]={0, 1, 0, 1, 0, 0, 0, 0, 0, 0} ;// e₁−e₉ elements are saved in e[1]-e[9]


    float w[]={0, 10, 9, 8, 7, 8, 6, 5, 4, 3, 2, 1} ;// w₁−w₁₀ elements are saved
                                                      in w[1]-w[10]
    float q[]={0, 1, 1, 0, 6, 5, 4, 3, 2, 1} ;// q₁−q₉ elements are saved in q[1]-q[9]
    float p[]={1, 0, 0, 0, 0, 0, 0, 0, 0, 0, 0} ;// initialize the characteristic polynomial
                                                    sequence p₀(x)−p₁₀(x)


float x=9; //      Assign x
int i, t=0, n=10; //   i is an intermediary variable, t represents the sign
                   change number, and n is the order of the matrix.
p[1]=x−w[1];
printf("\n p[0]=%f\n", p[0]);
printf("\n p[1]=%f\n", p[1]);
for(i=2;i<=n;i++)
{p[i]=(x−w[i])*p[i-1]−q[i-1]*e[i-1]*p[i-2];
printf("\n p[%d]=%f\n", i, p[i]);}
for(i=1;i<=n;i++)
if(p[i-1]*p[i]<0)t++;
else if(p[i]==0)
if(p[i-1]*p[i+1]<0)t++;
printf("\n t=%d\n", t);
system("pause");
 return 0;
}
```

See Fig. 2 for the program running results.

Put the statement float x = 9 in the program, replace with float x = 11. See Fig. 3 for the program running results.

Comprehensive analysis of 4.1–4.2 reveals that: As can be seen from Fig. 1–3, the matrix has ten eigenvalues of 8, 8.38, 10.61, 1, 2, 3, 4, 5, 6, 7.

$w_3, w_4, ..., w_{10} \notin [9, 11]$, $q_2 e_2 = q_3 e_3 = \cdots = q_9 e_9 = 0$, and $p_{10}(9) = -5040 \neq 0$, $p_{10}(11) = 544320 \neq 0$. t stands for sign change numbers of s(9) and s(11) respectively, and the difference between the two sign change numbers is s(a)-s(b) = s(9)-s(11) = 1–0 = 1, indicating that there are one eigenvalues in the interval [9], [11], which is 10.61. The program running result verifies that Theorem 1 holds.

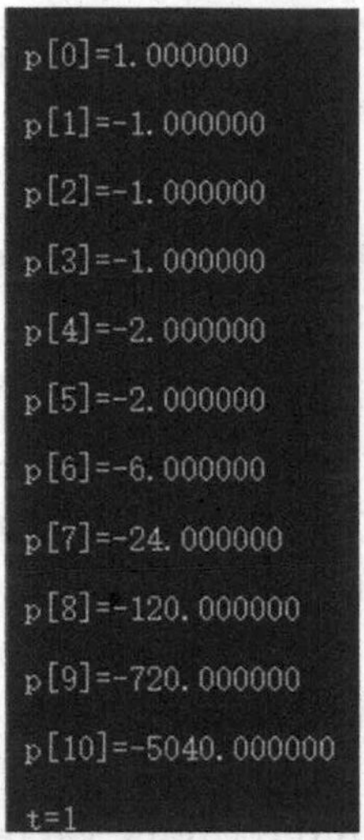

```
p[0]=1.000000
p[1]=-1.000000
p[2]=-1.000000
p[3]=-1.000000
p[4]=-2.000000
p[5]=-2.000000
p[6]=-6.000000
p[7]=-24.000000
p[8]=-120.000000
p[9]=-720.000000
p[10]=-5040.000000
t=1
```

Fig. 2. The program running results.

```
p[0]=1.000000
p[1]=1.000000
p[2]=1.000000
p[3]=3.000000
p[4]=12.000000
p[5]=36.000000
p[6]=180.000000
p[7]=1080.000000
p[8]=7560.000000
p[9]=60480.000000
p[10]=544320.000000
t=0
```

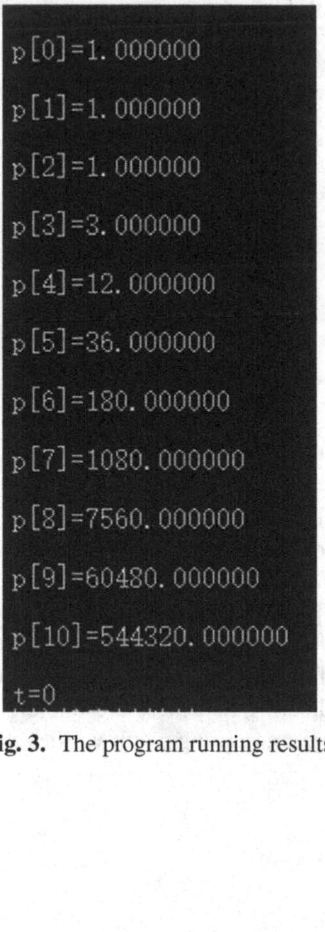

Fig. 3. The program running results.

5   Conclusions

T matrix has eight pairs of eigenvalues of adjacent two sequential principal square submatrices that are not strictly interlaced, that is, the matrix does not meet the key properties of using dichotomy to solve eigenvalues. It is pointed out that the dichotomy can be used to solve the eigenvalues of T matrices in interval $[a, b]$ when the $w_{n-7}, w_{n-6}, ..., w_n \notin [a, b]$ is satisfied. Although there are many algorithms to solve matrix eigenvalues, the research work in this paper not only expands the scope of application of dichotomy to solve matrix eigenvalues, but also provides a reference for expanding the scope of application of other algorithms to solve matrix eigenvalues.

6   Funding Projects

National Natural Science Foundation of China (11961021). Guangxi Natural Science Foundation of China (2020GXNSFAA159084). Guangxi Basic Research Capability Improvement Project for Young and Middle-aged University Teachers (2022KY0612). Hechi University Research Fund Project (2018XJQN007).

References

1. Wei, X., Ruidong, H., Haichao, L., et al.: Application of improved Fourier method in vibration characteristics analysis of beam structure. Noise Vib. Control **39**(1), 10–15 (2019)
2. Siyuan, B., Jinrui, C., Jing, Z.: Transverse vibration characteristics of nonlocal beams with arbitrary elastic boundary. J. Vib. Eng. **33**(4), 276–284 (2020)
3. Siyuan, B., Jing, Z., Jianwei, L.: Free vibration of multi-section beam with arbitrary elastic boundary. Appl. Math. Mech. **41**(9), 985–993 (2020)
4. Baocang, S., Lingmei, Z.: The eigenvalue solving problem of transfer matrix method. Noise Vib. Control (06), 34–35 (2007)
5. Fenglin, Y., Wenjun, M., Jie, Z., et al.: Calculation and comparison of critical force of step column model of telescopic boom crane. Mach. Des. Manuf. **5**(5), 23–27 (2020)
6. Haining, M., Kai, F., Lei, Z., Beibei, Z., Xinyu, T., Xinhong, H.: Short text clustering algorithm based on Laplacian atlas. Acta Electron. Sin. **49**(09), 1716–1723 (2021)
7. Jinfu, L.: Numerical Linear Algebra. Post &Telecom Press, Beijing (2006)
8. Xiaoguang, L., Xiaomei, L., Jianhua, C.: Dichotomy for solving the eigenvalue problem of symmetric banded matrix and its improvement. Comput. Phys. **14**(4–5), 450–452 (1997)
9. Erxiong, J.: Matrix Calculation. Science Press, Beijing (2008)
10. Zengxin, W., Weida, Q., Zhimei, Y.: Discussion on interval halving for solving symmetric triple-diagonal matrix eigenvalue. J. Chongqing Univ. Technol. (Nat. Sci. Edn.) **31**(1), 55–59 (2011)

Computer and Electronic Engineering

Federated Offline Reinforcement Learning for Autonomous Systems

Ju-eun Park and Honguk Woo[✉]

Department of Computer Science and Engineering, SungKyunKwan University,
Suwon, South Korea
{jueun.park,hwoo}@skku.edu

Abstract. Offline reinforcement learning (RL) provides a safe learning method that can be applied to real-world applications through a data-driven learning process. In general, this process attributes to learning on large datasets, which is similar to conventional supervised learning techniques. In this paper, we propose a federated offline RL framework where policy parameters learned locally on separate datasets of RL agents are aggregated in communication rounds to achieve a high-performance global policy. The framework can solve the data insufficiency issue by performing offline RL among multiple agents without data sharing and integration. We show that using the framework, agents achieve a stable-performance control policy in a 2D navigation environment, even when each has only 1% of the required training trajectory amount (e.g., 10 trajectories).

Keywords: Federated learning · Offline reinforcement learning · Autonomous systems

1 Introduction

Offline reinforcement learning (RL) enables policy learning on pre-collected and stored datasets without online interaction. It has been applied to automatic systems especially where interacting with environments is expensive or dangerous (e.g., self-driving, medicine, and robots). As the computing capacity of local devices has been improved, local data collection and local model training cases have increased. Accordingly, it is natural to consider offline RL as a local learning algorithm at individual devices for various autonomous decision-making scenarios. However, each device may have only a small dataset that is not sufficient to perform offline RL individually. Furthermore, each might not want to share its own data with others. That is, the so-called 'data island' phenomenon could frequently appear [1]. This phenomenon means that each dataset is collected separately by each RL agent without data sharing.

Figure 1 illustrates the performance degradation due to the lack of local datasets when offline RL is conducted. In the case of training an autonomous agent via offline RL (in a 2D navigation environment), the test scores of a policy learned on 1000 trajectories stably converge as the epoch increases, but the scores of a policy learned on relatively few trajectories (10 trajectories) are unstable. In this experiment, we used the conservative Q-learning algorithm (CQL) [2] for offline RL.

To address the data-insufficient issue in offline RL, in this work, we adopt a federated learning (FL) approach among multiple RL agents. Specifically, we present a federated offline RL framework in which the parameters of local policies are updated by individual agents and aggregated to obtain a global policy. Similar to FL in a conventional supervision setting, each agent does not share its own data with others, thereby being free from concerns about data privacy and regulation. Notice that offline RL is generally considered in areas such as medical care and autonomous driving, where it is difficult to learn policies through direct interaction with environments.

Through several experiments with the 2D navigation simulator, we evaluate the proposed framework, indicating the stable performance of RL policies achieved by the framework. For example, we show that the test scores of a global policy learned on a small dataset of 10 trajectories with 5 agents can reach about 96.14% of the performance of a policy learned on a large dataset of 1000 trajectories.

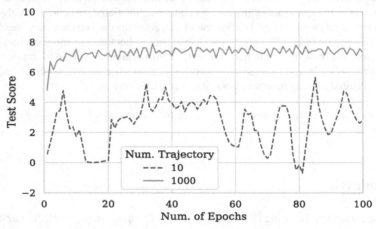

Fig. 1. The performance (test score) by offline RL (CQL) on the datasets of different trajectory sizes.

2 Federated Offline RL

In general, an RL agent is learned by continuously interacting with environments where a Markov decision process (MDP) is assumed. An MDP consists of state spaces $\mathcal{S}$, action spaces $\mathcal{A}$, transition probability p, and reward function r: $\{\mathcal{S}, \mathcal{A}, p, r\}$. In offline settings, an RL agent does not interact with environments directly for learning but only use offline buffers with datasets of pre-collected trajectories. Therefore, offline RL approaches have been considered beneficial under restrictive conditions where interaction with environments is at-risk or cost-expensive, e.g., dialog generation, robotic manipulation, and automatic navigation [3].

Different from conventional (online) RL approaches, the offline RL setting often causes the problem of distribution shift such that the learned policy (learning on the dataset) deviates from the behavior policy that generates the dataset [3]. Specifically, in the policy evaluation phase of off-policy actor-critic methods, the Q-function is trained on actions sampled from the behavior policy producing the dataset, while the target values use actions sampled from the learned policy. To treat the problem, several methods of regularizing the batch data or the Q-function have been studied. In [4], the batch regulation was suggested to minimize the extrapolation error that occurs due to the difference between the distribution of the offline dataset and the actual visited state-actions by the current learned policy. The agent named BCQ in that research takes actions similar to the offline dataset via regulation. BCQ employs the variational autoencoder (VAE) to generate plausible actions so that their distribution is similar to that of offline batches, rather than the actions derived from the learned policy.

A BCQ agent consists of a generative model $G_\omega(s) = \{E_{\omega_1}, D_{\omega_2}\}$, a perturbation model $\xi_\phi(s, a)$, and Q-networks $Q_{\theta_1}(s, a)$ and $Q_{\theta_2}(s, a)$. The Q-networks learning target of BCQ is given as

$$r + \gamma \max_{a_k} \left[\lambda \min_{j=1,2} Q_{\theta_j'}\left(s', a_k\right) + (1 - \lambda)\max_{j=1,2} Q_{\theta_j'}\left(s', a_k\right) \right] \tag{1}$$

where a_k is the perturbed actions sampled from $\xi_\phi(s, a)$, θ' is the target network, and λ is the soft clipped double Q-learning hyper-parameter.

In offline circumstances, the value overestimation problem by the distribution shift also occurs when off-policy RL methods are used. In [2], CQL intends not to overestimate the Q-value of the current learned policy by estimating its lower bound. The objective function of CQL is given as

$$\underset{Q}{\mathrm{argmin}} \alpha \mathbb{E}_{s \sim \mathcal{D}} \left[\log \sum_a \exp(Q(s, a)) - \mathbb{E}_{a \sim \hat{\pi}_\beta(a|s)}[Q(s, a)] \right] + \frac{1}{2} \mathbb{E}_{s,a,s' \sim \mathcal{D}} \left[\left(Q - \hat{\mathcal{B}}^{\pi^k} \hat{Q}^k \right)^2 \right] \tag{2}$$

where $Q(\cdot, \cdot)$ is the Q-function, $\mathcal{D} = \{\left(s, a, r, s'\right)\}$ is a dataset, $\hat{\pi}_\beta$ is the behavior policy, π^k is the learned policy at step k, and α is the tradeoff factor. $\hat{\mathcal{B}}$ is the single sample Bellman operator, i.e., $\hat{\mathcal{B}}^{\pi^k} \hat{Q}^k = r + \gamma \mathbb{E}_{a' \sim \pi^k(a'|s')}\left[\hat{Q}^k\left(s', a'\right)\right]$. This objective tends to reduce the biased error generated from out-of-distribution actions.

2.1 Federated Learning

Federated Learning (FL) is distributed machine learning technology in which the dataset locally owned by a client is not shared with a central server and other clients. FedAvg [5] is commonly used in practice, mostly for supervised learning models. In FedAvg, the averaging operation on the model parameters from multiple clients is conducted by a server at each communication (or federation) round, i.e.,

$$w' = \frac{1}{n}\sum_{i=0}^{n} w_i \tag{3}$$

where w_i is the local model parameter received from each client i, n is the number of clients who participated in the current communication round, and w' is the global model parameter for the next round. Each client receives w' and updates it to w_i by learning locally on its own dataset during each round. FL tends to be communication-efficient and privacy-preserving since only the model parameters are shared between clients and a server, not sharing datasets.

2.2 Federated Offline RL Procedure

Offline RL agents do not interact with an environment but use only their own datasets. In the case when such datasets are locally owned by different clients (where agents are trained) and cannot be integrated and shared, the policy performance by offline RL can be easily degraded due to insufficient quantity and low quality of a given dataset. We address this problem by employing FL algorithms for offline RL settings.

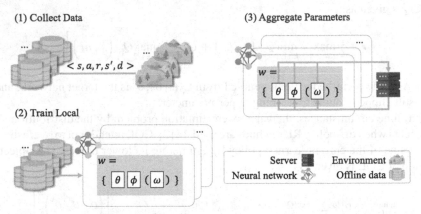

Fig. 2. Overall system concept. Datasets are collected from an environment. Each client locally trains its policy on the local dataset via the offline RL algorithm. The policy parameters are uploaded to the server after offline learning and aggregated by the federation algorithm. The aggregated parameters are then downloaded to clients, and the training procedure is repeated until convergence.

Figure 2 depicts the overall system, where n clients (offline RL agents) and a server engage in FL. The overall procedure includes (1) data collection, (2) local training, and (3) model aggregation phases. (1) Buffer data are gathered while a human policy or learned policy interacts with an environment. A pre-collected buffer contains a set of trajectory samples (s, a, r, s', d), where s and s' are the current state and the next state respectively, a is the action, r is the reward, and d is the Boolean which depicts whether the episode is completed. (2) An RL agent (client) is locally learned on its own buffer data using offline RL algorithms to update its policy parameters. (3) The RL agent sends the parameters to the server and the parameters received by the server are aggregated using Eq. (3) to update the global policy parameters. They are sent back to agents for the next rounds. This procedure is repeated until the converges global policy.

Algorithm 1: BCQ-FL Client

Input   : Number of FL round T_G, global model parameters $\mathbf{w} = \{\theta_g, \phi_g, \omega_g\}$, local client number i, dataset $\mathcal{D}_i$, client timestep T_C, target network update rate τ, mini-batch size N, max perturbation Φ and number of sampled actions n_a.

Output: Local parameters $\mathbf{w}^i = \{\theta^i, \phi^i, \omega^i\}$.

Initialize Q-networks $Q_{\theta_1}, Q_{\theta_2}$, perturbation network ξ_ϕ and VAE $G_\omega = \{E_{\omega_1}, D_{\omega_2}\}$, and target networks $Q_{\theta_1'}, Q_{\theta_2'}, \xi_\phi'$ with $\theta_1' \leftarrow \theta_1, \theta_2' \leftarrow \theta_2, \phi' \leftarrow \phi$.

for t in $\{1, \ldots, T_G\}$ **do**

    Receive global parameters $\mathbf{w}_{\theta,\phi,\omega}$ from the server.

    Set local parameters $\{\theta^i, \phi^i, \omega^i\} \leftarrow \mathbf{w}_{\theta,\phi,\omega}$.

    for t_c in $\{1, \ldots, T_C\}$ **do**

        Sample mini-batch of N transitions (s, a, r, s') from $\mathcal{D}_i$.

        $\mu, \sigma = E_{\omega_1^i}(s, a), \tilde{a} = D_{\omega_2^i}(s, z), z \sim \mathcal{N}(\mu, \sigma)$

        $\omega^i \leftarrow \arg\min_{\omega^i} \sum (a - \tilde{a})^2 + D_{KL}(\mathcal{N}(\mu, \sigma)\|\mathcal{N}(0, 1))$

        Sample n_a actions: $\{a_k \sim G_\omega(s')\}_{k=1}^{n_a}$

        Perturb each action: $\{a_k = a_k + \xi_{\phi^i}(s', a_k, \Phi)\}_{k=1}^{n_a}$

        Set value target y (Eq. (1)).

        $\theta^i \leftarrow \arg\min_{\theta^i} \sum (y - Q_{\theta^i}(s, a))^2$

        $\phi^i \leftarrow \arg\max_{\phi^i} \sum Q_{\theta_1^i}(s, a + \xi_{\phi^i}(s, a, \Phi)), a \sim G_{\omega^i}(s)$

        Update target networks: $\theta_j'^i \leftarrow \tau\theta^i + (1 - \tau)\theta_j'^i, \phi'^i \leftarrow \tau\phi^i + (1 - \tau)\phi'^i$

    end

    Send $\mathbf{w}^i = \{\theta^i, \phi^i, \omega^i\}$ to the server.

end

Algorithm 2: CQL-FL Client

Input   : Number of FL round T_G, global model parameters $\mathbf{w} = \{\theta_g, \phi_g\}$, Q-network and policy network gradient steps G_Q and G_π, Q-network and policy network learning rate η_Q and η_π, local client number i, dataset $\mathcal{D}_i$ and client timestep T_C.

Output: Local parameters $\mathbf{w}^i = \{\theta^i, \phi^i\}$.

Initialize Q-network Q_θ and a policy network π_ϕ.

for t in $\{1, \ldots, T_G\}$ **do**

    Receive global parameters $\mathbf{w}_{\theta,\phi}$ from the server.

    Set local parameters: $\{\theta^i, \phi^i\} \leftarrow \mathbf{w}_{\theta,\phi}$

    for t_c in $\{1, \ldots, T_C\}$ **do**

        Train the Q_{θ^i} on objective from Eq. (2) using G_Q steps and $\mathcal{D}_i$:

        $\theta_t^i \leftarrow \theta_{t-1}^i - \eta_Q \nabla_{\theta^i} CQL(\mathcal{H})(\theta^i)$ using Bellman operator $\mathcal{B}^{\pi_{\phi_t^i}}$.

        Improve policy π_{ϕ^i} with SAC-style entropy regularization using G_π steps and $\mathcal{D}_i$: $\phi_t^i \leftarrow \phi_{t-1}^i + \eta_\pi \mathbb{E}_{s \sim \mathcal{D}_i, a \sim \pi_{\phi^i}(\cdot|s)} [Q_{\theta^i}(s, a) - \log\pi_{\phi^i}(a|s)]$

    end

    Send $\mathbf{w}^i = \{\theta^i, \phi^i\}$ to the server.

end

Algorithm 3: FedAvg Server

Input : Number of FL round T_G, parameter list Π, number of clients
　　　　n, local parameters $\{\mathbf{w}^i\}_{i=0}^n$
Output: Global parameters $\mathbf{w}$
Initialize global parameters $\mathbf{w}$.
for t in $\{1, \ldots, T_G\}$ **do**
　　for i in $\{1, \ldots, n\}$ **do**
　　　　Send $\mathbf{w}$ to client i.
　　　　Receive $\mathbf{w}^i$ from client i.
　　　　Append $\mathbf{w}^i$ to Π.
　　end
　　$\mathbf{w} = \mathtt{FedAvg}(\Pi)$ (Eq. (3))
end

Algorithms 1 and 2 show the federated offline RL process of the client-side with BCQ [4] and CQL [2] algorithms, respectively. The BCQ policy network is composed of three parameter sets including Q-network parameters θ, perturbation network parameters ϕ, and generative model network parameters ω. Similarly, the SAC [6] based CQL contains Q-network parameters θ and policy parameters ϕ. Algorithm 3 depicts the parameter aggregation procedure at the server-side of FL, working with both Algorithms 1 and 2, where sending and receiving parameters can be executed asynchronously.

3   Evaluation

To implement the proposed framework, we use Python version 3.7, `flwr` library [7], and `d3rlpy` [8]. `flwr` is applied for the FL procedure between offline RL clients and a FedAvg server, and `d3rlpy` is used in the client-side offline RL algorithms, BCQ and CQL. The hyper-parameter settings are listed in Table 1.

We use a 2D-navigation environment to collect datasets where simple operations are simulated for the 2D version of autonomous navigation systems. The target task in this environment is set to drive a drone to reach a goal position as fast as possible while avoiding moving obstacles. The start position is set to the center of the given square map and the goal position is set to its bottom left.

The position of multiple moving obstacles can be detected through 2D LiDAR sensors of the drone. Accordingly, the observations (RL states) contain 360° 2D LiDAR sensor values in addition to several other measurements such as drone position, energy, velocity, and distance to the goal. The actions are given as the velocity setting for x- and y-coordinates. The reward of each step is given based on the difference between the previous and current distance to the goal position. When the drone collides with moving obstacles or reaches the goal position, the episode is terminated. In addition, the test score signifies the cumulative reward and is used as the performance evaluation metric. Moreover, we collect only 10 trajectories for each client because we assume the data-insufficient settings.

Table 1. Hyper-parameters.

Algorithm	Hyper-parameter	Value
BCQ & CQL	Discount factor	0.9900
BCQ	Actor learning rate	0.0010
	Actor optimizer	Adam
	Critic learning rate	0.0010
	Critic optimizer	Adam
	Imitator (VAE) learning rate	0.0010
	Imitator optimizer	Adam
	Batch size 100 Target update rate (τ)	0.0050
	Constrained region parameter (Φ)	0.0500
	Critic ensemble weight factor (λ)	0.7500
CQL	Actor learning rate (η_π)	0.0001
	Actor optimizer	Adam
	Critic learning rate (η_Q)	0.0003
	Critic optimizer	Adam
	SAC temperature learning rate	0.0001
	SAC temperature optimizer	Adam
	Initial temperature	1
	Tradeoff factor (α) learning rate	0.0001
	α optimizer	Adam
	Initial alpha	1
	α threshold	10
	Batch size	256
	Target update rate (τ)	0.0050
	Conservative weight	5
	The number of action samples	10

3.1 Federated Offline RL Performance

Figure 3 depicts the performance via FL over communication rounds achieved by two different offline RL algorithms, CQL-FL (Algorithm 2) and BCQ-FL (Algorithm 1), respectively. The number of clients in this experiment is set to 5 and the number of local epochs is set to 5. As expected, the test score (the cumulative reward in an episode) of both algorithms increases as more communication rounds are given. Interestingly, CQL-FL converges faster than BCQ-FL, whilst their final scores are almost the same. The numerical results are shown in Table 2, where the performance of federated offline RL is compared with (non-federated, local) offline RL trained on a relatively large

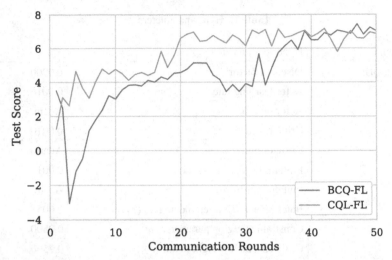

Fig. 3. Performance of federated offline RL over communication rounds.

dataset (i.e., 1000 trajectories). Note that the offline RL score graph was previously discussed in Fig. 1. We set the final score of conventional offline RL (CQL) learned on 1000 trajectories to 100% and compare it with CQL-FL and BCQ-FL. Importantly, both BCQ-FL and CQL-FL learned on only a small number of trajectories (10 for each client) perform comparatively to CQL learned on a large number of trajectories (1000); i.e., with 10 trajectories and 5 clients, BCQ-FL achieves 96.14% and CQL-FL archives 93.83%, compared to CQL with 1000 trajectories.

Table 2. Evaluation summary.

Method	CQL (Offline RL)	BCQ-FL	CQL-FL
Num. trajectories	1000	10	10
Final test score	7.31	7.03	6.86
Final test score std.	2.28	2.69	2.83
Relative score	100.00%	96.14%	93.83%

We also evaluate the effect of the number of local epochs E in Fig. 4. Obviously, the small epoch settings show relatively poor performance, while about 50 epochs achieve stable performance. In this experiment, we test E = 1, 5, 10, 20, 50, 100 with the default condition of 50 communication rounds and 5 clients.

In Fig. 5, we evaluate the performance with respect to various FL group sizes. It is expected that more clients in the FL group are likely to improve the final performance, as more clients might be seen as more data used for FL to some extent in homogeneous settings. A group of 2 clients achieves poor performance, while CQL-FL with 3, 4, and 5 clients shows relatively stable performance, and also BCQ-FL with 4 and 5 clients

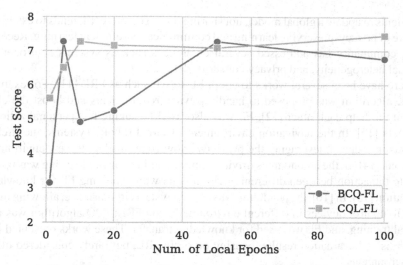

Fig. 4. Performance with respect to the number of local epochs.

shows stable performance. In this experiment, we test the number of clients $K = 2, 3,$ 4, 5 with the default condition of 50 communication rounds and 5 local epochs.

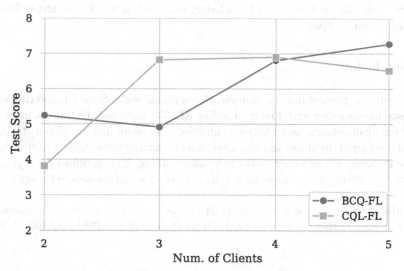

Fig. 5. Performance with respect to the number of clients.

4   Related Works

Distributed machine learning approaches without sharing local data have been discussed in the FL context [5, 9]. In each round, the FL server aggregates the model parameters from clients to update a global model, not sharing data. Therefore, FL tends to establish a structure of privacy-preserving learning and communication-efficient learning. Recently, the FL community has addressed several challenges such as systems heterogeneity, statistical heterogeneity, and privacy concerns [10].

There have been several works to adopt FL to RL such as FRL [11]. For example, an FRL algorithm was proposed to handle privacy requirements and boost the client capability to help each other [12]. FL was also used for taking fast personalization of RL agents [13]. In the navigation environment of cloud robotic systems, the lifelong FRL was introduced, leveraging the previous knowledge to adapt to new environment conditions [14]. In the autonomous driving domain, online transfer learning was used to translate the action between different environments while applying FL for knowledge aggregation [15]. In [16], the pendulum was set in physical circumstance, allowing multi-agents to control devices in different environments via FRL; PPO algorithm was used for local training, and FL was used for knowledge transfer. These works presented FRL algorithms and evaluation results focusing on online RL, but rarely considered offline RL environments.

There exists a gap between an optimal policy and a policy learned on pre-collected datasets of offline RL. To tackle the gap, batch constrained Q-learning (BCQ) [4] regulates the actions of a learned policy using restrictive perturbation, and conservative Q-learning (CQL) [2] regulates the Q-value by estimating its lower bound. In this work, we adopt these offline RL algorithms to FL, and evaluate their performance in various FL conditions. To the best of our knowledge, our work is the first to test offline RL algorithms with FL clients.

5   Conclusion

In this work, we presented an FL framework integrated with offline RL algorithms and addressed the data-insufficient issue of offline RL. We evaluated the framework performance on small datasets under various conditions, showing that the global RL policy learned through FL in offline learning environments can achieve stable performance.

The direction to our future works is to adapt offline RL algorithms for a group of agents with statistically heterogeneous behavior policies and datasets in FL settings.

Acknowledgements. This work was supported by the Institute for Information and Communications Technology Planning and Evaluation (IITP) under Grant 2021-0-00875 and 2021-0-00900.

References

1. Zhang, C., Xie, Y., Bai, H., Yu, B., Li, W., Gao, Y.: A survey on federated learning. Knowl.-Based Syst. **216**, 106775 (2021)
2. Kumar, A., Zhou, A., Tucker, G., Levine, S.: Conservative q-learning for offline reinforcement learning. arXiv preprint arXiv:2006.04779 (2020)
3. Levine, S., Kumar, A., Tucker, G., Fu, J.: Offline reinforcement learning: tutorial, review, and perspectives on open problems. arXiv preprint arXiv:2005.01643 (2020)

4. Fujimoto, S., Meger, D., Precup, D.: Off-policy deep reinforcement learning without exploration. In: ICML 2019, pp. 2052–2062. PMLR (2019)
5. McMahan, B., Moore, E., Ramage, D., Hampson, S., Arcas, B.A.: Communication-efficient learning of deep networks from decentralized data. In: AISTATS 2017, vol. 54, pp. 1273–1282. PMLR (2017)
6. Haarnoja, T., Zhou, A., Abbeel, P., Levine, S.: Soft actor-critic: off-policy maximum entropy deep reinforcement learning with a stochastic actor. In: International Conference on Machine Learning, pp. 1861–1870. PMLR (2018)
7. Beutel, D.J., Topal, T., Mathur, A., Qiu, X., Parcollet, T., Lane, N.D.: Flower: a friendly federated learning research framework. arXiv preprint arXiv:2007.14390 (2020)
8. Takuma Seno, M.I.: d3rlpy: an offline deep reinforcement library. In: NeurIPS 2021 Offline Reinforcement Learning Workshop (2021)
9. Konečny, J., McMahan, H.B., Yu, F.X., Richtárik, P., Suresh, A.T., Bacon, D.: Federated learning: strategies for improving communication efficiency. arXiv preprint arXiv:1610.05492 (2016)
10. Li, T., Sahu, A.K., Talwalkar, A., Smith, V.: Federated learning: challenges, methods, and future directions. IEEE Signal Process. Mag. 37(3), 50–60 (2020)
11. Qi, J., Zhou, Q., Lei, L., Zheng, K.: Federated reinforcement learning: techniques, applications, and open challenges. arXiv preprint arXiv:2108.11887 (2021)
12. Zhuo, H.H., Feng, W., Xu, Q., Yang, Q., Lin, Y.: Federated reinforcement learning. arXiv preprint arXiv:1901.08277 (2019)
13. Nadiger, C., Kumar, A., Abdelhak, S.: Federated reinforcement learning for fast personalization. In: 2019 AIKE, pp. 123–127. IEEE (2019)
14. Liu, B., Wang, L., Liu, M.: Lifelong federated reinforcement learning: a learning architecture for navigation in cloud robotic systems. IEEE Robot. Autom. Lett. 4(4), 4555–4562 (2019)
15. Liang, X., Liu, Y., Chen, T., Liu, M., Yang, Q.: Federated transfer reinforcement learning for autonomous driving. arXiv preprint arXiv:1910.06001 (2019)
16. Lim, H.K., Kim, J.B., Kim, C.M., Hwang, G.Y., Choi, H.B., Han, Y.H.: Federated reinforcement learning for controlling multiple rotary inverted pendulums in edge computing environments. In: ICAIIC 2020, pp. 463–464. IEEE (2020)

A New Smart Controller Model Using Feedback and Feedforward for Three Phase Inverter

Mohammad A. Obeidat[1]($\boxtimes$) (iD), Osama Al Smeerat[2], and Ayman M. Mansour[3] (iD)

[1] Department of Electrical Engineering, College of Engineering,
Al-Ahliyya Amman University, Amman, Jordan
m.obeidat@ammanu.edu.jo
[2] Electrical Distribution Company (EDCO), Amman, Jordan
[3] Department of Computer and Communications Engineering, College of Engineering,
Tafila Technical University, Tafila 66110, Jordan

Abstract. This paper aims to discussing three-phase inverter design and Control using feedback and feedforward controllers. These inverters convert direct current to sinusoidal voltages and currents in three phases. The behavior of feedback, feedforward controller, is addressed in the presence of DC source input disturbances. These disturbances cause by the direct current source (PV, batteries, etc.). After conducting the study and simulation, it establishes that the feedback controller is ineffective at stabilizing the system when the input disturbances happen; hence, the feedforward controller is used to achieve this objective. This review aims to establish a stable and controllable approach for a three-phase voltage source inverter that supplies the most frequently used applications (RL load) and investigates the behavior of feedback and feedforward controllers. Finally, the paper developed an optimization method for improving the performance of a realistic inverter under any load condition.

Keywords: Three phase inverter · Feedback controller · Feedforward controller · Smart controller

1 Introduction

Power converters use to convert alternating current (AC) to direct current. David Prince wrote an article entitled "The Inverter" It was supposed to be equipment that acts as a rectifier but in a reversed method, hence an inverter. Only after the invention of the thyristors in 1950 that the inverter and rectifier started to achieve their highest possible level of productivity [1].

The development of pulse width modulated (PWM) converters has helped promote small-scale DC power sources that provide power in direct current. However, to supplement the traditional grid, the DC/AC power converters must transfer the delivered power from these DC sources to AC sources [2]. The study of power electronics entails examining electronic circuits designed to regulate the flow of electrical energy. These circuits can withstand even more power than the individual devices are rate to [3].

F. Neri et al. (Eds.): CCCE 2022, CCIS 1630, pp. 118–126, 2022.
https://doi.org/10.1007/978-3-031-17422-3_11

The voltage-source inverter drive technology use in many diverse applications such as UPS (Uninterrupted Power Supply), AC machines, and AC battery-powered devices.in standard Outputs according to the switching modes, we have square wave used in some low-performance systems; also, we can use PWM or a Sinusoidal signal for AC or DC drives [4].

Also, the voltage source inverter has uses in renewable energy; the real worry about higher energy costs, the exhaustion of conventional fuels, and the recent adoption of clean energy sources for generating electricity have encouraged the industry to produce more renewable electricity.

Additionally, the cost of PV cells became feasible where the wind turbines became sophisticated; unlike typical sinusoidal electricity, renewable sources do not generate conventional sinusoidal voltage, and the current cannot be stored, therefore inverter used in these systems to convert the output into AC waveform to connect them to a grid or to keep current in DC batteries.

Based on the definition of the linear system, we can automatically define the nonlinear system as the system which breaks the principle of superposition. The system, which has parameters changing with time, belongs to the time-varying system; otherwise, the system will be a time-invariant system.

Control can be found in many applications in our life such as an automobile with the position of the paddle as input and the speed as output, a traffic light whose input is the color of the light and production is the traffic flow, and the bank accounts with fund deposited as input and the interest generated as the output [5–7].

Two main types of control systems are considering in applying controllers to the systems [8, 9]. An open-loop system controls its outputs without needing a feedback loop. This method does not allow a comparison between the input and the output; therefore, this system is also known as Non-Feedback System.

The output should be driven ideally by the reference signal, and that could be possible only if the controller has been designed based on the complete knowledge of the plant. However, if any change in the plant happened, caused by a disturbance occurrence, the controller in the open-loop will not be effective anymore, and errors will emerge.

The closed-loop control system uses feedback to adjust the system's output. Since the output and input have been comparing, we have a system that may make errors known as an error signal. In such system, there is a connection between the output and the controller. Any disturbance that causes the plant to change would result in a rise in the error value on the controller side. Thus, the controller can take action to compensate for the disturbance. The problem that has been demonstrating in the open-lope scheme will be solved using the closed-loop structure.

This control system expects it to have an optimum output that serves the system's primary function for any scenario. If we conclude that dynamic performance behavior is critical, finding it must be a significant part of any scheme. The placement of the controller, which needs in any device, depends on the user's commands. We need to define the relationship between the plant and the controller in any controlled system, using an open-loop control system or a closed-loop control system [5–7].

A proportional-integral-derivative controller (PID controller) is commonly using as a feedback mechanism in industrial control systems. It utilizes a PID controller to identify a mismatch between a measured value and a target value calculated by a set point. It tries to correct the error by using a controllable variable.

PID controller consists of three terms; proportional to the current value of the error (P), the term which reflects the past value of the error using the integrator (I), and the estimator term of the future value of the error using the deviation (D). All PID controller terms can tune by using the changeable gain (K) [8, 9].

They presented a multi-path feedforward controller designed in the discrete-time domain for a three-phase inverter with a step-up transformer. They tested their model under resistive inductive and nonlinear loads. The findings indicate that the proposed model enhances the steady-state and dynamic behavior of the system. The technique suggested based on the assumption that it would reduce system impedance to avoid voltage drop [10].

Introduced a detailed description of a simple feedforward approach to stabilize the three-phase voltage source inverter system, which fed squirrel cage induction motor with LC filter by subtracting the feedforward term from d-q current component overcome the resonant that produce between the LC filter and rotor–flux oriented Control. The result shows that the stability analysis of the drive with Feedforward validates at different set points, and the overall system became more efficient [11].

The effectiveness of the implementation of voltage feedforward on-grid tide inverter has adversely affected the voltage feedforward in the weak grid. The author proposed proportional voltage feedforward to improve system stability and power quality also eliminate the harmonics in the grid by adding proportional coefficient 'k' in the feedforward path. The Theories and actual findings confirm the efficacy of the system.

It is common knowledge that specific switching devices operate between the source and load. The amount of switching devices is constrained by the complexity, though, so we can only choose the simplest. Also, the most complicated circuit has at least one switch as an input/output link. A converter has a certain amount of 'n' inputs and a certain amount of 'm' outputs, so the number of switching devices needed for electrical energy conversion is equal to 'm × n' [12].

Also, for example, to understand conversion from single-phase to DC, suppose we have three terminals on the single-phase side (input), and the DC has three terminals (output); thus, a total of '3 × 3 = 9' switches are required.

2   The Developed Feedback and Feedforward Model

We have various controller structures and will implement the proportional-integral (PI) controller, which will be using in our design.

The following formula can use to formulate the PI controller:

$$PI(s) = K1 + \frac{K2}{S}$$

where:

K1: represents the proportional gain, K2: represents the integral gain.

The FB structure shows in Fig. 1 based on the structure, we must select the gains K1 and K2 to modify the system's dynamic response; in this case, we can use the PID tuning MATLAB software determine those gains. The configuration of the closed-loop control system would be as follows:

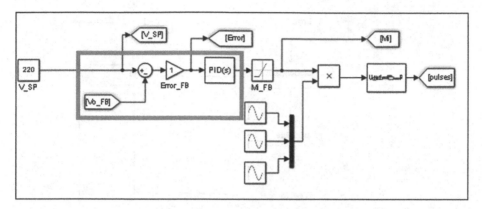

Fig. 1. The FB structure for voltage source inverter.

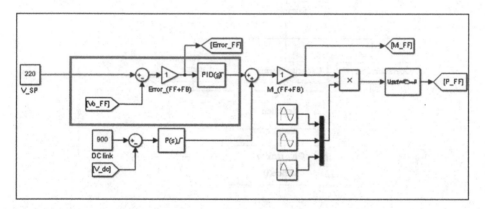

Fig. 2. The voltage source inverter FB and FF structures.

When we were using just the feedback controller, the input disturbances cannot be modified when the input contains abrupt jumps. Consequently, the input should be controllable to modify in response to the input's sudden jumps. This aim was meeting by the implementation of a feedforward controller into the system. Thus the system will continuously update the input variance and proactive in taking action before these variation signals appear on the output. As shown in Fig. 2, the structure that provides us with this method is called the feedforward (FF) structure.

To calculate the operation cost of different stations to feed all the loads with inverter, the system will then be ascending order the stations depending on the cost by using fuzzy logic control [13]. Then it chooses the station which have lower operating cost. After arranging the stations, the station will be automatically loaded and the system will sense if the load is greater than the capacity of the selected station or not, if yes load sharing will do to feed the reminder of load all this will be by using fuzzy logic control Fig. 3 shows chart flow for principle of work of the intelligent model.

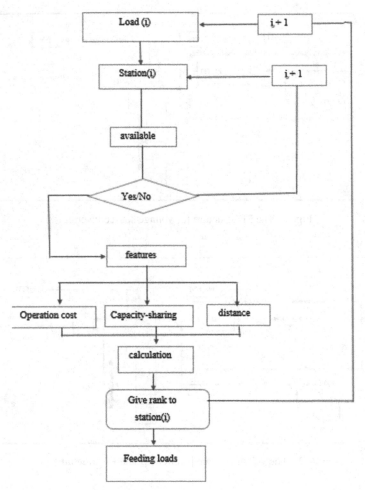

Fig. 3. Chart flow for intelligent model.

3   Results and Discussion

This chapter introduced and explained the achieved results of the feedback and feedforward/feedback controller's responses under various DC link voltage disturbances and load conditions. The disturbance signal was divided into three parts as shown below in Fig. 4 first positive edge, first negative edge, and second positive edge, where the response of the two controllers to be studied was recorded, and the readings were then analyzed accordingly.

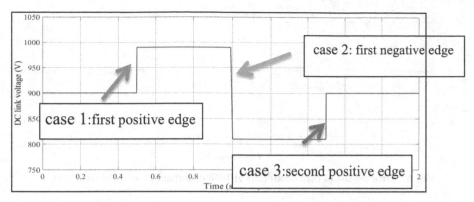

Fig. 4. DC link disturbance shows three disturbance cases.

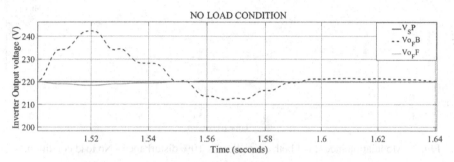

Fig. 5. Output voltage responses at 10% disturbance - No load condition.

In this case, the following parameters of VSI circuit shown in will be used, then the disturbance in DC link input by -10% around the set point will be applied. The second positive edge analysis from Figs. 5, 6, 7 and 8 and Table 1 showed that load increment has a positive effect on the rise time and the settling time, and at the same time it has a negative effect on the overshoot values. It is also noticeable that the proposed control system scored the minimum overshoot value, which means the generation of a fixed and stable voltage signal.

The developed system can be use in many applications such as electrical vehicles, vehicle to vehicle-to-vehicle communication systems, prediction systems, and energy management systems [14–17].

4 Conclusion

This paper proposes an optimal feedforward controller for compensating for the abrupt changes in the input voltage of a three-phase voltage source inverter. The results indicate that, due to the inherently unstable existence of DC sources such as a photovoltaic device or a battery, the feedback controller cannot tolerate input jumps. The monitoring and noise rejection shortcomings of the proposed system with the Feedback controller can be addressed only by adding Feedforward to the system. The feedforward scheme represents

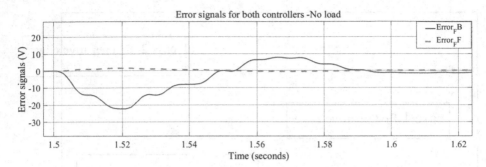

Fig. 6. Error signals of both controllers at 10% disturbance - No load condition.

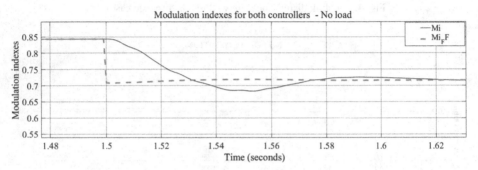

Fig. 7. Modulation indexes of both controllers at 10% disturbance - No load condition.

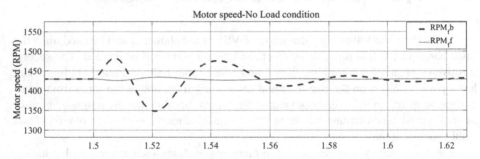

Fig. 8. Motor speed of both controllers at 10% disturbance - No load condition.

good performance and stabilizes the system, especially in the first negative edge; also, in the second positive edge. The overall system overshot and rise time becomes more amelioration with the feedforward controller; furthermore, significantly improves the accuracy of fundamental voltage and current components and the measurement efficiency of inverter simulations.

Table 1. Second positive edge analysis with 10% disturbance.

Comparison criterion	No load measurements		Full load measurements	
	FB	FF + FB	FB	FF + FB
Disturbance	**10%**		**10%**	
Rise time	2.69E-05	2.45E-05	2.64E-04	3.42E-04
Settling time	0.1928	0.1990	0.1789	0.1820
Settling Min	212.1462	218.4726	213.7508	218.7520
Settling Max	242.3014	220.5213	243.6854	220.3296
Over shoot	22.3014	0.5213	23.6854	0.3296
Under shoot	7.8538	1.5274	6.2492	1.2480
Peak	242.3014	220.5213	243.6854	220.3296
Peak time	0.0464	0.0990	0.0340	0.0865

References

1. Owen, E.L.: History [origin of the inverter]. IEEE Ind. Appl. Mag. **2**(1), 64–66 (1996). https://doi.org/10.1109/2943.476602
2. Prince, D.C.: "The Inverter," GE Review, vol. 28, no. 10, pp. 676–681 (1925). A Sarkar- 2015 -repository.asu.edu, Modeling and Control of a Three Phase Voltage Source Inverter with an LCL Filter
3. Llorente, R.M.: Practical control of electric machines. In: Advances in Industrial Control, Springer, Cham (2020). https://doi.org/10.1007/978-3-030-34758-1
4. Boeing, G.: Visual analysis of nonlinear dynamical systems: chaos, fractals, self-similarity and the limits of prediction. Systems **4**(4), 37 (2016). https://doi.org/10.3390/systems4040037
5. Ogata, K.: Discrete-Time Control Systems. Prentice-Hall, Upper Saddle River, NJ (1987)
6. Lin, F.: Robust Control Design. Wiley, Chichester (2007). Araki, M. "PID Control" (PDF)
7. https://www.motioncontroltips.com/faq-what-are-pid-gains-and-feed-forward-gains/
8. Bowei, L., Li, P., Ke, X.: A multi path feedforward control of load current for three-phase inverter with transformer. In: 2019 IEEE 28th International Symposium (2019)
9. Mishra, P., Maheshwari, R.: A simple feed-forward approach to stabilize VSI-fed induction motor with filter in RFOC. In: IEEE Transactions on Industrial Electronics (2019)
10. Wang, G., Du, X., Sun, P., Tai, H.-M., Ji, Y.: Analysis and design of voltage feedforward for stability and power quality of grid-tied inverter. In: IECON 2017 - 43rd Annual Conference of the IEEE Industrial Electronics Society (2017)
11. Liu, T., Hao, X., Yang, X., Zhao, M., Xiong, L.S.: A novel grid voltage feedforward control strategy for grid-connected inverters in weak grid. In: IECON 2017 - 43rd Annual Conference of the IEEE Industrial Electronics Society (2017). https://doi.org/10.1109/iecon.2017.821 6109
12. Cho, H.Y., Santi, E.: Modeling and stability analysis in multi-converter systems including positive feedforward control. In: 2008 34th Annual Conference of IEEE Industrial Electronics (2008). https://doi.org/10.1109/iecon.2008.4758062
13. Obeidat, M.A., Qawaqneh, M., Mansour, A.M., Abdallah, J.: Smart distribution system using fuzzy logic control. In: 2021 12th International Renewable Engineering IEEE Conference (IREC), pp. 1–5 (2021). https://doi.org/10.1109/IREC51415.2021.9427799

14. Obeidat, M.A., et al.: Effect of electric vehicles charging loads on realistic residential distribution system in Aqaba-Jordan. World Electr. Veh. J. **12**(4), 218 (2021). https://doi.org/10.3390/wevj12040218
15. Mansour, A.M., Almutairi, A., Alyami, S., Obeidat, M.A., Almkahles, D., Sathik, J.: A unique unified wind speed approach to decision-making for dispersed locations. Sustainability **13**(16), 9340 (2021). https://doi.org/10.3390/su13169340
16. Obeidat, M.A., Al Anani, M., Mansour, A.M.: Online parameter estimation of DC-DC converter through OPC communication channel. Int. J. Adv. Comput. Sci. Appl. **12**(5), 102–114 (2021). https://doi.org/10.14569/IJACSA.2021.0120514
17. Obeidat, M.A.: Real-time DC servomotor identification and control of mechanical braking system for vehicle to vehicle communication. Int. J. Comput. Appl. **182**(40), 20–30 (2019)

Adaptive Design Space Reconstruction Method in Surrogate Based Global Optimization

Yingtao Zuo[1], Chao Wang[2], Wei Zhang[1($\boxtimes$)], Lu Xia[1], and Zhenghong Gao[1]

[1] School of Aeronautics, Northwestern Polytechnical University, Xi'an 710072, China
gemini@mail.nwpu.edu.cn

[2] Beijing Institute of Electronic System Engineer, Beijing 100000, People's Republic of China

Abstract. Surrogate-based global optimization (SBO) has gained rapid dominance in engineering design. However, traditional SBO method over entire design space with large size interval would be considerably time-consuming. In order to improve the optimization efficiency in SBO, an adaptive design space reconstruction (ADS) method based on fuzzy clustering method and effective sample points is proposed in this paper. Fuzzy c mean clustering method is applied to divide the initial design space into several sub-regions from which we choose the sub-region which is most likely to contain the global optima. During the optimization process, effective sample points are collected to be the center of new space constructed by trust region method, instead of a single sample point, to keep optimization from getting trapped in local minimums. Then the optimization search will be managed in the reconstructed promising sub-region. We test and verify the proposed method with the airfoil drag minimization problems proposed by Aerodynamic Design Optimization Discussion Group (ADODG), which could demonstrate that better results can be obtained within the reconstructed design space with high efficiency.

1 Introduction

Due to the increasingly complexity of engineering design, surrogate-based global optimization (SBO) methods are widely used to reduce the computational cost [1–3]. However, the size interval of design space has large influence on accuracy of surrogate model, so as to the efficiency of optimization. Too small the design space is, it would bear the risks that the global optima could be outside the current design space; As the design space becomes larger, it will bring burden to the accuracy of surrogate, leading the failure on finding the global optima. To solve these problems, an effective branch of research is the design space reduction (DSR).

The basic idea of DSR is to gradually resize the design space during the optimization process. In general, the DSR could be classified into two categories: (1) reducing the number of design variables, which is also called dimensionality reduction. Satyajit et al. [4] presented an innovative proper orthogonal decomposition based reduced order design scheme method to reduce the number of design variable. Steven et al. [5] used principal component analysis to reduce space dimensionality based on the covariance matrix of the gradient. Trent et al. [6] employed the active subspace method to exploit the low

F. Neri et al. (Eds.): CCCE 2022, CCIS 1630, pp. 127–138, 2022.
https://doi.org/10.1007/978-3-031-17422-3_12

dimensionality space with an evenly spread set of observed sample points. Asha et al. [7] transformed the high dimensionality data set into a low dimensional latent space by nonlinear latent variables model called generative topographic mapping. Nowadays, the nonlinear dimensionality reduction methods have become more and more popular for its ability on capturing the feature of latent space [8]; (2) reducing the size interval of the design variables. The widen of design space brings significant computational burden to the improve approximation level of surrogate model. To solve this problem, the designers usually tend to define conservative bound to design variables based on their prior knowledge, which would lack applicability for other design problems. To get rid of this situation, plenty of researches on design space reduction have been developed. Wang et al. [9] utilized the proposed fuzzy clustering based hierarchical metamodeling to intuitively capture promising regions and efficiently identify the near-global design optima. Tseng et al. [10] proposed a novel design space reduction method to shrink the space range rapidly before the simulation-based local search started, which could decrease computational costs without sacrificing accuracy. Yong Wang et al. [11] focused search effort onto specific area of feasible region by shrinking the constrained search space. Long et al. [12] developed a trust region sampling space method to reduce the space range gradually. The sequential sampling as well as adaptive surrogate method are used to improve the optimization efficiency and convergence. These methods could reduce large initial design space into a small near-optima region, while the new sub-region based on only one sample point could lead premature of optimization [9]. Besides, most of these methods treat all design variables uniformly when reducing the size interval, which is not sufficient for problems with high dimensionality.

To shake off the limitation of design space's range, reduce the computational cost and improve the efficiency of design space exploration, this paper proposes an adaptive design space reconstruction based on the fuzzy c mean clustering and effective sample points, drawing the initial space into a relatively small near-optima space. The method can be roughly divided into four parts: (1) obtain the reduced the design space using the FCM; (2) add new sample point to sample library to update the kriging model; (3) collect the effective sample points to reconstruct the design space; (4) resize the range of the sensitive design variable if necessary.

The rest of this paper is as follows. Related information of fuzzy c mean clustering, and trust region method are presented in Sect. 2. In the Sect. 3, the proposed methodology is introduced. In Sect. 4, the proposed method is applied to two airfoil optimization problems. Finally, the concluding remarks and future work are given.

2 Background

2.1 Fuzzy C Mean Clustering

Fuzzy c mean (FCM) is a data clustering technique originally introduced by Bezdek [13]. The FCM algorithm is one of the most widely used clustering algorithms in engineering design [9].

The objective of FCM is to minimize the cost function formulated as Eq. (1),

$$J(U, V) = \sum_{j=1}^{c} \sum_{i=1}^{N} (\mu_{ij})^m \|x_i - v_j\|^2 \tag{1}$$

where $V = \{v_1, v_2, ...v_c\}$ represents the cluster centers, N is the number of sample points and c is the number of clusters. The exponent $m \in [1, \infty]$ (typically $m = 2$) is a weighting factor to measuring the fuzziness of each cluster. $U = (\mu_{ij})_{N*c}$ is a fuzzy partition matrix, where the μ_{ij} reflects the degree of membership between the x_i and the jth cluster:

$$\mu_{ij} = \frac{1}{\sum_{k=1}^{c} (\|x_i - v_j\| / \|x_i - v_k\|)^{\frac{2}{m-1}}} \tag{2}$$

The value of matrix U should satisfy the following conditions: $\sum_{j=1}^{c} \mu_{ij} = 1$, $0 \le \mu_{ij} \le 1$. For a specified number of clusters, given the exponent m and the termination criteria, $\max_{ij}\{|\mu_{ij}^{(T+1)} - \mu_{ij}^{(T)}|\} < \varepsilon$, $\varepsilon \in [0, 1]$ [0, 1]. The optimal solution U_* and V_* is obtained on the following conditions:

$$Min\{J_{cm}(U, V)\} \tag{3}$$

The global optima would be more possible to be nearby the cluster with better objective values. Even if the optima locate outside the initial design space, better sample points will be gained in fewer iterations and the possibility of finding the global optima will be greatly increased.

In this paper, the fuzzy c mean clustering is adopted for its simplicity, robustness and convenience. But the proposed method does not dictate the exclusive use of fuzzy c mean clustering while other clustering methods may be equally acceptable.

2.2 Trust Region Method

The trust region method (TRM) [14, 15] is a classical space reduction sequential sampling methods. It is used to construct the promising sampling space step by step based on current best point and fitting quality of the surrogate model. In TRM, the current best sample point is chosen as the center of new design space. And the trust factor r_k and trust region radius δ_k are computed through Eq. (4) and Eq. (5):

$$r_k = \frac{f(x_{k-1}) - f(x_k)}{\tilde{f}(x_{k-1}) - \tilde{f}(x_k)} = \frac{\Delta f}{\tilde{f}_{k-1} - \tilde{f}_k} \tag{4}$$

$$\delta_k = \begin{cases} c_1 \|x_k - x_{k-1}\| & r_k < r_1 \\ \min(c_2 \|x_k - x_{k-1}\|, \Delta) & r_k > r_2 \\ \|x_k - x_{k-1}\| & r_2 \ge r_k \ge r_1 \\ \delta_{min} & r_k < 0 \end{cases} \tag{5}$$

In this paper, "~" means surrogate prediction. If $\Delta f > 0$, the current surrogate model is capable of searching better design result, and then the trust radius δ_k is updated according to Eq. (5), in which $c_1 = 0.75$, $c_2 = 1.25$, $r_1 = 0.1$, $r_2 = 0.75$ [12]. Otherwise, the current surrogate has bad performance and plenty of computational cost would be need to find the global optima. The TRM is then forced to exploit a small neighborhood around the current best point. Although the TRM has shown improvements in optimization efficiency, it may lead the optimization to get trapped in local minimums and be premature for it use only one point as the center at one time.

3 Proposed Method

In the optimization process, the functions (both objective functions and constraint functions) usually guide the optimization from the initial broad design space into a small region nearby the global optima. With the increment of size interval, it would impose daunting computational cost. To efficiently reduce the size interval of design space, this paper proposes an adaptive design space reconstruction method to resize the design space. The flowchart of the method is illustrated in Fig. 1, and the following are the detailed steps of this method:

Step 1: *generate initial sample points and construct Kriging model.* At the beginning step, the initial sample points are generated by the Latin Hypercube sampling (LHS) [16] method, sampling the design space more uniformly and can achieve better approximation with fewer sample points.

Step 2: *clustering and space determined.* In this step, the FCM is utilized to cluster the found sample points. The number of clusters is usually set two as discussed in [9]. Usually there are some properties can be used to choose the ideal cluster such as mean, variance, maximum and minimum of sample points. In this paper we make the constraint (both geometry and aerodynamic constraints) one of the properties to be considered in choosing the cluster. The cluster outside the constraint boundary is got rid when constructing the design space. To decide how much each constraint function must be expanded in order to provide the sufficient number of sample points to construct the design space during clustering, we consider the concept of ε tubes from the support vector regression [17], which could also be used on collecting effective sample points.

Step 3: *identify the reconstruction moment.* The moment to reconstruct the design space is hard to identify, as it is difficult to differentiate whether optimizer converging to optima or the optimizer traversing a complicated area of the design space [18]. In this paper, we define the moment that objective function value keeps unchanged for N iterations during the exploration of current design space. The small value for N would lead a more aggressive system. For complex optimization problem, this can cause premature when the objective function value only makes a small improvement through highly non-linear area. For this reason, N can be increased for exploration and only trigger reconstruction when objective function value consistently stagnates.

Step 4: *select the effective sample points.* Sample points with good performance are defined as the effective sample points. The space based on effective sample points describes a landscape near the global optima, whose variation can reflect the trend of optimization. That's to say, as effective sample points update, the reconstructed sub-region would approach the global optima step by step. During this process, the quantity of effective sample points directly influences the size of the sub-region. Too large the number is, it will bring burden to exploration in new sub-region because of the bad surrogate accuracy. Therefore, we retain the best 20% of sample points within the constraints boundary as the effective sample points [7]. At the beginning, the difference between the effective sample points would make sure the reconstructed sub-region contains global optima. After a few iterations, the difference of each effective sample point could grow smaller, the reconstructed sub-region shrinks significantly, resulting in high efficiency of optimization.

Step 5: *reconstruct the sub-region.* The sub-region is not only determined by the boundary of the collected effective sample points. To guarantee an adequate design space, here we make use of the trust region method (TRM), which resizes new design space in the light of the improvement of objective function value. The space is adaptively adjusted by the center and radius. In the proposed method, when the trigger moment comes, the effective sample points are selected as the center of new space. L_k is the unit radius defined as the 10% of size interval on each dimensionality, where $VR\,max$ and $VR\,min$ are the upper and lower bound of the new design space. And radius factor c is decided by surrogate accuracy threshold Tr and rolling average of surrogate error $\widetilde{fr}$. The trust radius r_k is updated according to Eq. (8) where the typical values of c_1, c_2 are used, namely, $c_1 = 0.5$, $c_2 = 2$. In Eq. (7), $c_2 * L_k$ is the upper bound of the radius.

$$L_k = 0.1 * (VR\,max - VR\,min) \tag{6}$$

$$c = \frac{Tr}{\sum\limits_{i=k-N+1}^{k} \widetilde{fr}_i} \tag{7}$$

$$r_k = \begin{cases} c * L_k & c_1 \le c \le c_2 \\ \min(c * L_k, c_2 * L_k) & c_2 < c \\ 0 & c \le c_1 \end{cases} \tag{8}$$

It's well known that for surrogate-based global optimization, surrogate accuracy is related to the sample density of design space. Sometimes, the reconstructed sub-region may far beyond the boundary of the initial design space leading the local sample points distribution extremely sparse. Ordinarily, there're two ways to improve the local sample density, one is to generate sample points in the current design space which would lead to extra computational cost. The other one is to shrink the space size to guarantee the surrogate accuracy when surrogate has a bad performance. However, each dimensionality of design space has different impact on the objective function even if they vary in a same range. Dimensionalities making dramatic changes of objective function manipulate bigger design space from the view of parameterization, meaning more sensitive to objective function. Therefore, there's no need to narrow all dimensionalities if we can figure out sensitive design variables. To distinguish these design variables, we adopt the elementary effect method. In this method, the large mean indicates a latent dimensionality with large influence on the objective function, whereas large variance indicates latent dimensionalities responsible for nonlinear effects. Let $\Psi(q)$ be defined as

$$\Psi(q) = \frac{\sum_{j \in J(q)} v_j}{\sum_{j=1}^{p} v_j} \geq c \tag{9}$$

where $0 \leq c \leq 1$ is some fraction of the total sensitivity. For example, $J(q)$ might be selected such that $\Psi(q) \geq 0.8$ to capture 80% of the sensitivity, which are defined as main effect variables.

4   Test on Benchmark Problems

In this section, the proposed ADS method was tested on the airfoil optimization cases provided by ADODG. Moreover, it was compared to the multi-round optimization method with fixed design space, to test and verify its global convergence, efficiency and robustness.

4.1   Case I. Symmetric Transonic Airfoil Design

The first test case involves drag minimization for a symmetric airfoil, NACA 0012, under inviscid transonic condition. The design Mach number is 0.85, while the angle of attack is fixed at $\alpha = 0^{o}$. Since the airfoil in this case should be symmetric, the upper half with a symmetry boundary condition is used. Under this circumstance, the constraints satisfied when $y \geq y_{baseline}$ everywhere on the upper surface. The problem can be summarized as

$$\begin{aligned} Min \quad & C_d \\ s.t. \quad & t \geq t_{baseline} \; \forall x \in [0, 1] \end{aligned} \tag{10}$$

From the reference [19–21] we can know there exists great shape difference between the optimized foil and traditional airfoils, which requires an unconventional design space to get to the optima. It will test the applicability of proposed method on transferring the design space to find the ideal design result.

The Bezier curve with 24 design variables is adopt to deform the airfoil and the initial design space is shown as Fig. 2, where 100 initial sample points are generated by LHS method. Table 1 and Fig. 3 demonstrate the set of the computational grid. Another 200 sample points would be infilled during optimization process.

The optimized airfoils and corresponding pressure distribution are presented in Fig. 4 and Fig. 5, respectively. It's shown that all optimized airfoils share the similar shape deformation, which means the optimization tendency is efficient for the proposed method. These are strong shocks extending far into the flow field on the initial geometry. Due to the thickness constraints, the optimizer thickens the airfoil and create a surface that delays the pressure recovery, leading weaker shocks to occur near the trailing edge of the optimized airfoil. The later pressure recovers, the smaller drag coefficient it occurs.

Figure 6 shows the cost effectiveness comparisons of adaptive design space method (ADS) and fixed design space (FDS), where FDS round 2 is based on the FDS round 1 as multi-round optimization. It turns out the ADS strongly outperforms FDS. Optimization

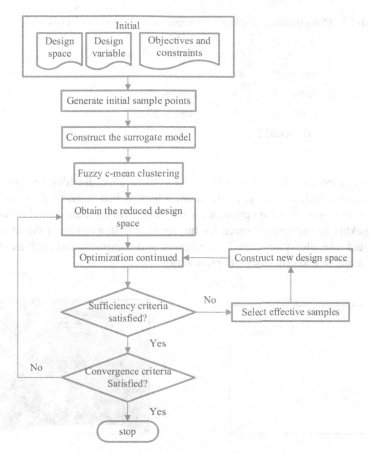

Fig. 1. Flowchart of method.

Table 1. Summary of grid's properties.

Parameter	Size
Far field	50
Grid size	201 × 201
Off wall spacing	5e−4
Growth ratio	1.05
Leading edge spacing	0.001
Trailing edge spacing	0.001

in the fixed design space stalls quite early, indicating that space size interval is the limitation that keeps the optimizer from reaching the global optima in this case. Figure 7 shows the final space size interval of ADS, which has large difference with the initial

Table 2. Computational results of optimized airfoils by different design space.

	C_d (cts)	Computational cost (CFD)
Baseline	471	
ADS	34	100 + 200
FDS round 1	174	100 + 200
FDS round 2	56	100 + 200

design space, especially for the regions on the trailing edge. Besides, the size interval of final adaptive design space is really small which is beneficial to fitting quality of surrogate model, as well as the optimization efficiency. As for multi-round optimization method, making up the enough space for further exploration to obtain the ideal design results would take plenty of work for designers and computational cost, as shown in Table 2, leading to bad efficiency of optimization.

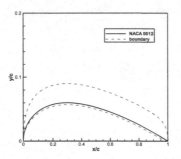

Fig. 2. Initial design space of NACA0012 airfoil.

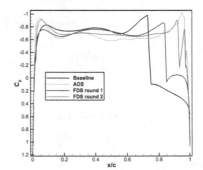

Fig. 3. The computational grid of NACA0012.

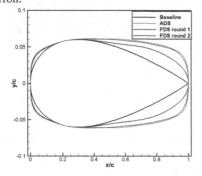

Fig. 4. Comparison of airfoil shape

Fig. 5. Comparison of pressure distribution

4.2 Case II. Transonic Airfoil Design

The Case II revisits transonic airfoil design (Mach 0.734). The objective is to reduce the drag coefficient, while constraints are imposed on lift coefficient, pitching moment

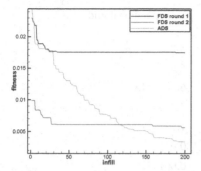

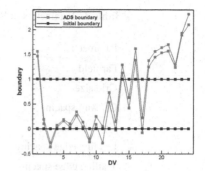

Fig. 6. Cost-effectiveness of different design space

Fig. 7. Boundaries of different design space

coefficient and the area, as shown in Eq. (11):

$$Min \quad C_d$$
$$s.t. \quad C_l = 0.824$$
$$|C_m| \leq 0.092 \tag{11}$$
$$Area \geq Area_{initial}$$

The baseline shape is the RAE 2822 airfoil. The design variables are the z-coordinates of the CST parameterization with 24 shape parameters in total. The initial design space is shown as Fig. 8. Similar to Case I, 100 sample points are generated in the initial design space with LHS method, and 200 sample points are to be infilled by EI method during optimization progress. The computational grid (Fig. 9) setting is shown as Table 3.

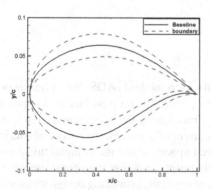

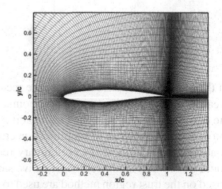

Fig. 8. Initial design space

Fig. 9. The computational grid of RAE 2822

The comparisons of optimized airfoils and their pressure distribution are shown in Fig. 10 and Fig. 11, respectively. From Fig. 11 we can know that the suction peak of the optimized airfoil gets much stronger and the shock wave is nearly eliminated, compared to RAE 2822 airfoil. The leading edges of all optimized airfoils get sharper, and upper surfaces become more flat, which is good for pressure recovery and shock wave elimination. The comparisons of cost-effectiveness of different optimization methods are

Table 3. Parameters of RAE 2822 computational grid.

Parameter	Size
Far field	50
Grid size	601×213
Off wall spacing	5e−6
Growth ratio	1.13
Leading edge spacing	0.001
Trailing edge spacing	0.001

shown in Fig. 12, showing that the result of ADS and FDS round 2 are much better than that of FDS round 1. The optimizer in FDS round 1 stall so early as a result of the boundary limitation (Fig. 13). With larger design space, FDS round 2 acquires a better result eventually, which is similar to ADS referring to Table 4.

Table 4. Computational results of optimized airfoils by different design space.

	C_d (cts)	C_m	Area	Computational cost (CFD)
RAE2822	203.1	−0.0927	0.07787	
ADS	110.6	−0.0918	0.07787	100 + 200
FDS round 1	112.1	−0.0908	0.07787	100 + 200
FDS round 2	111.1	−0.0917	0.07787	100 + 200

5 Conclusion

In this paper, an adaptive design space reconstruction method (ADS) based on fuzzy c mean clustering and effective sample points is proposed. The proposed method improves the optimization efficiency and applicability by transferring the inefficient design space into a sub-region where the optimization can be more efficient. Fuzzy c mean clustering is adopted to determine the preliminary reduced space so that the computational cost could be reduced significantly. Effective sample points selection with constraints rules based on the trust region method are used to reconstruct the promising design space with the help of elementary effect method. The performance of the proposed method is tested on two airfoil optimization problems compared to the multi-round optimization method. The comparison results reveal that the proposed method gains better performance on global convergence, efficiency and robustness. Nowadays, with the development of aviation engineering demands, the aerodynamic shape of modern aircrafts will be largely distinct from the traditional ones. Under these circumstances, ADS exhibits a good prospect to efficiently solve modern unconventional aircraft design problems. However, the current work still has some limitations, as follows. Firstly, fuzzy c mean clustering is not good enough when the sample distribution is not the convex for it's based on Euclidean distance. Secondly, ADS may be still trapped in local optima when solving some complex functions with massive and crowded local optima with the limitation of trust region method. Although effective sample points selection could ease that situation, the selection rules could still be consummate. Further enhancement of ADS is expected to upgrade the global exploration.

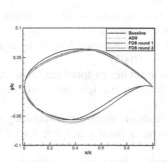

Fig. 10. Comparison of airfoil shape.

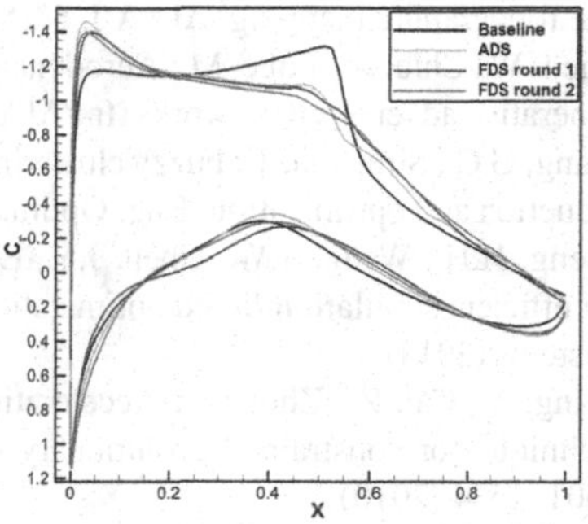

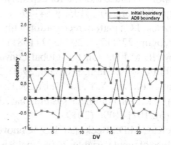

Fig. 11. Comparison of pressure distribution.

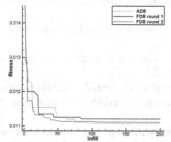

Fig. 12. Cost-effectiveness of different design space.

Fig. 13. Boundary of different design space.

Acknowledgement. The authors would like to acknowledge the financial support received from the key laboratory funding with the reference number 6142201200106 and natural science funding with the reference number 11772266.

References

1. Queipo, N.V., Haftka, R.T., Wei, S., Goel, T., Vaidyanathan, R., Tucker, P.K.: Surrogate-based analysis and optimization. Prog. Aerosp. Sci. **41**(1), 1–28 (2005)
2. Fernández-Godino, M.G., Haftka, R.T., Balachandar, S., Gogu, C., Bartoli, N., Dubreuil, S.: Noise filtering and uncertainty quantification in surrogate based optimization. In: 2018 AIAA Non-Deterministic Approaches Conference (2018)
3. Soilahoudine, M., Gogu, C., Bes, C.: Accelerated adaptive surrogate-based optimization through reduced-order modeling. AIAA J. **55**(5), 1681–1694 (2017). https://doi.org/10.2514/1.j055252

4. Ghoman, S., Wang, Z., Ping, C., Kapania, R.: A POD-based reduced order design scheme for shape optimization of air vehicles. In: AIAA/ASME/ASCE/AHS/ASC Structures, Structural Dynamics & Materials Conference AIAA/ASME/AHS Adaptive Structures Conference AIAA (2013)
5. Berguin, S.H., Mavris, D.N.: Dimensionality reduction using principal component analysis applied to the gradient. AIAA J. **53**(4), 1078–1090 (2014)
6. Capristan, F.M., Alonso, J.J.: Active subspaces applied to range safety analysis and optimization. In: 17th AIAA Non-Deterministic Approaches Conference (2015)
7. Viswanath, A., Forrester, A.I.J., Keane, A.J.: Constrained design optimization using generative topographic mapping. AIAA J. **52**(5), 1010–1023 (2014)
8. Chen, W., Chiu, K., Fuge, M.: Aerodynamic design optimization and shape exploration using generative adversarial networks. In: AIAA Scitech 2019 Forum (2019)
9. Wang, G.G., Simpson, T.: Fuzzy clustering based hierarchical metamodeling for design space reduction and optimization. Eng. Optim. **36**(3), 313–335 (2004)
10. Tseng, H.H., Wang, S.W., Chen, J.Y., Liu, C.N.J.: A novel design space reduction method for efficient simulation-based optimization. In: IEEE International Symposium on Circuits & Systems (2014)
11. Wang, Y., Cai, Z., Zhou, Y.: Accelerating adaptive trade-off model using shrinking space technique for constrained evolutionary optimization. Int. J. Numer. Methods Eng. **77**(11), 1501–1534 (2010)
12. Long, T., Li, X., Shi, R., Liu, J., Guo, X., Liu, L.: Gradient-free trust-region-based adaptive response surface method for expensive aircraft optimization. AIAA J. **56**(2), 862–873 (2018). https://doi.org/10.2514/1.j054779
13. Bezdek, J.C.: Pattern recognition with fuzzy objective function algorithms. Adv. Appl. Pattern Recognit. **22**(1171), 203–239 (1981)
14. Powell, M.J.D.: On the global convergence of trust region algorithms for unconstrained minimization. Math. Program. **29**(3), 297–303 (1984)
15. Sun, Z.B., Sun, Y.Y., Li, Y., Liu, K.P.: A new trust region–sequential quadratic programming approach for nonlinear systems based on nonlinear model predictive control. Eng. Optim. **51**(6), 1071–1096 (2019)
16. McKay, M.D., Beckman, R.J., Conover, W.J.: A comparison of three methods for selecting values of input variables in the analysis of output from a computer code. Technometrics **42**(1), 55–61 (2000)
17. Chen, P.H., Lin, C.J., Scholkopf, B.: A tutorial on v-support vector machines. Appl. Stoch. Models Bus. Ind. **21**(2), 111–136 (2005)
18. Masters, D.A., Taylor, N.J., Rendall, T., Allen, C.B.: Progressive subdivision curves for aerodynamic shape optimisation. In: 54th AIAA Aerospace Sciences Meeting (2016)
19. Nadarajah, S.: Adjoint-based aerodynamic optimization of benchmark problems. In: 53rd AIAA Aerospace Sciences Meeting (2015)
20. Poole, D.J., Allen, C.B., Rendall, T.: Control point-based aerodynamic shape optimization applied to AIAA ADODG test cases. AIAA J. (2015)
21. Carrier, G., et al.: Gradient-based aerodynamic optimization with the elsA software. In: 52nd Aerospace Sciences Meeting - AIAA Scitech (2014)

Research on Ship Carbon Emission Statistical Optimization Based on MRV Rules

Pu Chen, Hailin Zheng$^{(\boxtimes)}$, and Bin Wu

School of Naval Architecture and Maritime, Zhejiang Ocean University, Zhoushan 316022,
China
hlzhzjou@126.com

Abstract. Transportation, which accounts for 20% of the world's energy consumption and greenhouse gas emissions, is still rising fast. How to reduce greenhouse gas emissions and reduce the energy consumption of ships has become a major issue. This paper compares and analyzes four monitoring methods: fuel supply tracking, fuel tank level monitoring, flow meter monitoring and carbon dioxide gas monitoring. Carbon emission sources such as main engine, auxiliary engine, boiler, gas turbine and inert gas generator are monitored. Then, the quantitative methods of ship carbon emissions are compared, and the influence of ship impact factors on the calculation results is analyzed, so as to give the optimization measures.

Keywords: Ship · Carbon emission source · Monitoring method · Statistical optimization

1 Introduction

On March 22, 2018, the international energy agency (IEA) released its 2017 global energy and carbon dioxide status report, in which it announced that global energy-related carbon dioxide (CO_2) emissions (the largest source of man-made greenhouse gas emissions) increased 1.4% year on year to 32.5 billion tons in 2017, a record high. In the global carbon emissions, the carbon emissions of ships account for the largest proportion. With the increase of the carbon emissions of ships, the pollution of ships to the environment has also been widely concerned by the society. The research on ship carbon emission monitoring methods mainly involves the specific interests of stakeholders such as ports, shipowners and other relevant shipping departments in economy, health and environmental protection. It is of positive significance in reducing the adverse impact of air pollution on human environment, and also relates to the realization of emission reduction targets of operating ships. Therefore, this paper studies the current monitoring methods and carbon emission sources of ship carbon emissions and proposes corresponding optimization measures [1].

F. Neri et al. (Eds.): CCCE 2022, CCIS 1630, pp. 139–147, 2022.
https://doi.org/10.1007/978-3-031-17422-3_13

2   Carbon Emission Monitoring Methods for Ships

Carbon emissions from ships sailing on the sea mainly come from main engines, auxiliaries, boilers, gas turbines and inert gas generators. The carbon emission sources use different types of fuel, so the carbon emission factors are not the same. Ship fuel oil is generally divided into two types, namely residue type (heavy oil) and distillate type (light oil). In addition, many ships will use cargo as fuel, such as LNG ships.

For long-haul cargo ships, the main engine type is generally a low-speed diesel engine (two-stroke), using heavy oil, corresponding to the carbon emission factor of 3.114 or 3.151, see Table 1 for details.

Table 1.   Characteristics of each carbon source of the ship.

Source of carbon emission	Machine type	Fuel type	Emission factor
Main engine	Low speed diesel engine	Heavy fuel oil	3.114
		Light fuel oil	3.151
Auxiliary engine	Medium/high speed diesel engine	Light fuel oil	3.206
Boiler	Auxiliary boiler, or Medium/high speed diesel engine	Light fuel oil	3.206

2.1   Fuel Supply List (BDN) Tracking and Bunker Periodic Inventory

BDN is provided by the fuel supplier as follows: 1) receiving the name of the fuel supply vessel and its IMO number; 2) fuel supply port; 3) fuel supply start time; 4) fuel supplier name, address and telephone; 5) fuel Name; 6) quantity of fuel (metric tons); 7) density of fuel at 15 degrees Celsius (kg/m^3); 8) sulphur content of fuel (%m/m); 9) a statement signed and certified by the supplier's representative (Prove that the fuel supplied is in accordance with paragraphs 14.1 or 4(a) and 18 of Annex 6 of the MARPOL Convention).

The calculation method of fuel consumption per unit time is as follows:

Fuel consumption is equal to the amount of fuel at the beginning of the period plus the amount of BDN minus the amount of fuel at the end of the period and then minus the amount of fuel at the end of the period.

Due to its poor accuracy and limited application range, this method cannot analyze each carbon emission source of ships one by one, and can only roughly calculate the total fuel consumption of carbon emission source, and it must be used in combination with the fuel tank level monitoring method.

2.2   Fuel Tank Level Monitoring

This method is simple and relatively low in cost. It is also unable to monitor various carbon emission sources of ships. The monitoring frequency is generally 2 times a day, and 1 time every 15 min when refueling. The liquid level is monitored by sounding equipment to read the level height of the fuel tank, which is converted into fuel volume after measurement, and then converted into fuel weight according to the fuel density (which can be obtained through BDN).

The calculation method of the fuel consumption of a ship in unit time or fixed voyage is as follows:

(Navigation on the sea) Fuel consumption equals to level in the tank when leaving the current port of call plus level difference when refueling during the voyage minus level in the tank when arriving at the next adjacent port minus level when the fuel is rejected during the voyage difference (parking) fuel consumption equals to tank level at the current port of call plus level difference when refueling during port stop minus level in tank when leaving the current port of call minus level difference when pumping out fuel during port stop.

2.3 Flow Meter Monitoring for Fuel Combustion Process

The electronic flow meter mainly measures the volume of the accumulative flow to monitor the carbon emission source of the ship main engine. The volumetric flow meter is mainly used for carbon emission sources such as auxiliary engines and boilers driven by medium-high speed diesel engines. The accuracy is up to 0.1–0.2%.

The types and characteristics of the flow meter are shown in Table 2:

Table 2. Flow meter types and characteristics for fuel monitoring.

Category	Subcategory	Measurement parameter	Accuracy	Applicable object or place
Electronic flow meters	/	Cumulative flow (volume)	0.2%	Main engine
Velocity sensing flow meters	Turbine meter	Instantaneous flow rate (volume of liquid flowing per unit time)	N/A	Large ship
Inferential flow meters	Variable aperture meter	Hydraulic difference	3.0%	/
Optical flow meters	N/A	Instantaneous flow rate (volume of liquid flowing per unit time)	N/A	/
Positive displacement flow meters	Oval gear, rotary piston	Cumulative flow (volume)	0.1–0.2%	High speed fluid
Mass sensing flow meters	Coriolis meters	Instantaneous flow rate (quality of liquid flowing per unit time)	0.05–0.2%	High value fluid

Fuel consumption per unit time or fixed voyage is calculated as follows:

(Sea voyage) Fuel consumption is equal to the sum of the flow meters at each carbon emission source during the voyage.

(Docking port) Fuel consumption is equal to the sum of the flow meters at each carbon emission source during the port stay.

2.4 Direct Carbon Emission Measurement

There are four kinds of methods for ship carbon emission monitoring, seen in Table 3 for details. The use of waste gas flow meter, through the main engine, auxiliary machinery, boiler chimney and other places directly measure the ship's carbon emissions, high accuracy, high cost, and for all types of carbon emission sources are applicable.

Fuel consumption per unit time or fixed voyage is calculated as follows:

(Sea voyage) Fuel consumption is equal to the total exhaust gas during the voyage times the ratio of the co_2 concentration to the carbon emission factor.

(Docking port) Fuel consumption is equal to the total amount of exhaust gas during port stay times the ratio of co_2 concentration to carbon emission factor.

3 Carbon Emission Calculation Method of Ship

3.1 Carbon Emission Calculation Method Based on IMO Emission Factor

In 2008, IMO MEPC/57 conference put forward new shipbuilding aiming to reduce ship emissions of greenhouse gases "CO_2 design index" (EEDI) and operation of the ship operating efficiency index (EEOI), involving two index calculation of fuel CO_2 conversion coefficient of emission factors, namely the EEOI unit transport power is defined as the ship amount of CO_2 emissions, MEPC. 1/Circ. 684 circular the vessel operating efficiency index (EEOI) voluntary use guide recommended EEOI formula (1):

$$EEOI = \frac{\sum_{j} FC_j \times CF_j}{m_{c\,arg\,o} \times D} \tag{1}$$

where FC_j means the actual consumption of fuel of class j, CF_j means the carbon dioxide emission factor, $m_{c\,arg\,o}$ means the cargo transportation volume and D means the cargo transportation distance [2–7].

3.2 Calculation Method for Monitoring Oil Consumption of Flow Meter

The fuel consumption of the ship mainly comes from the main engine and secondary engine oil consumption. The fuel consumption of ship is directly proportional to the cubic power of ship speed [8]. Fuel consumption times carbon emission factor is equal to carbon emissions, the carbon emissions of a single voyage can be calculated by formula (2).

$$E_{CX} = EF_{CX} \times \left[MF_K \times \left(\frac{S_{1k}}{S_{0k}} \right) + AF_K \right] \times \frac{d_{ij}}{24S_{1k}} \tag{2}$$

where E_{CX} means the total carbon emissions of a ship for a voyage, EF_{CX} means fuel consumption, i means the starting port, j means the arrival port, MF_K means the daily fuel consumption of the main engine, S_{1k} means the real-time speed, S_{0k} means the rated speed, AF_K means the daily fuel consumption of the auxiliary machine, d_{ij} means the distance between the ports.

Table 3. Comparative analysis of ship carbon emission monitoring methods

	Costs/burden for ship owner or operator	Accuracy	Verification cost	Monitoring emissions types	Voyage monitoring	Annual monitoring	Emissions from each carbon source	Feedback timeliness	Mandatory	Data Consistency	Applicable to carbon emission sources
BDN tracking	No equipping cost	1–5%	highest	CO_2, SOX	Not applicable	applicable	Not clear	Serious lag behind	yes	Difficult to ensure	Main engine, auxiliary engine and boiler (cannot be monitored separately)
Tank sounding	1000–3000 USD	2–5%	higher	CO_2, SOX	applicable	applicable	clear	lag behind	no	Difficult to ensure	Main engine, auxiliary engine and boiler (cannot be monitored separately)
Flow meter monitoring	15000–60000 USD	0.05–2%	low	CO_2, SOX	applicable	applicable	clear	Real-time	no	Easier to ensure	Electronic flowmeter (main engine), volumetric flowmeter (auxiliary boiler)
Direct emission monitoring	100000 USD	2%	low	CO_2, SOX, NOX, PM and so on	applicable	applicable	clear	Real-time	no	Easier to ensure	Main engine, auxiliary engine, boiler (can be monitored separately)

3.3 Direct Carbon Emission Measurement Calculation Method

In unit time or fixed voyage, the exhaust gas flow meter is used to measure the concentration of carbon dioxide, the temperature of flue gas, the volume percentage of water vapor in the flue gas, the static pressure of flue gas and the flow rate of flue gas. An enhanced flue gas analyzer and a temperature and humidity meter are needed. Firstly, the carbon dioxide emission rate is calculated according to the clapeyron equation [9], as follows:

$$M_{CO2} = \frac{P * V_{总} \times (1 - \varphi) \times 44}{R * T} \times (V_{CO2} + V_{CO}) \times 10^{-6} \tag{3}$$

where M_{CO2} indicates the instantaneous emission rate (t/s) of carbon dioxide, P means the absolute pressure (pa) at the flue measurement point, $V_{总}$ means the total volume of the flue at the flue monitoring point (m3/s), and the monitoring of the flue gas. V_{CO2} and V_{CO} means the volume fraction (%) of carbon dioxide and carbon monoxide, R means the standard gas molar volume, T means the thermodynamic temperature (K).

1) The calculation method of carbon dioxide emissions during the monitoring period is as follows:

$$S_{CO2} = M_{CO2} * t \tag{4}$$

2) Calculate the carbon emissions by the mass ratio method, that is, calculate the carbon dioxide consumption for a period of time by the ratio of the fuel consumption to the measured time, as in Eq. (5):

$$S_m = \frac{S_{CO2} * N_m}{N} \tag{5}$$

where S_m indicates the amount of carbon emissions(t) in the measured time, N_m indicates the fuel consumption(t) during the recording period, N indicates the fuel consumption(t) during the monitoring period.

3) The time ratio method calculates the carbon dioxide emissions, that is, calculates the carbon emissions by the ratio of the sailing time to the time period, as shown in Eq. (6):

$$S_m = \frac{S_{CO2} \times t_m}{t} \tag{6}$$

4) The load ratio method calculates the carbon dioxide emissions, that is, the carbon emission rate and the ship load are used to calculate the carbon emissions, as in Eq. (7):

$$S_m = \frac{M_{CO2} \times G_m \times t_m}{F} \tag{7}$$

where F indicates that the equipment monitors the average load (t/h), G_m indicates that the equipment is normally negative (t/h).

4   Analysis of Existing Problems in Carbon Emission Statistics of Water Transport Ships

4.1   The Monitoring Method Has a Large Error to the Statistical Results

Fuel supply note (BDN) tracking and bunker periodic inventory of vessels not suitable for cargo fuel or not available to BDN and must be used in conjunction with bunker level monitoring methods. The calculated fuel consumption of ships is larger than the actual one because the fuel in fuel piping system is neglected. Fuel tank level monitoring can be carried out manually or electronically, but the accuracy varies with the ship structure and software changes, and the calculation results do not include the fuel remaining in the fuel relationship, so the calculation results are larger; The flow meter is used to monitor the fuel combustion process in the fuel equipment. The calculated result is closer to the actual fuel consumed by the ship, and it is easy to distinguish the carbon emission of the ship within and outside the EU region, which is convenient for the preparation of carbon emission report, but the monitoring equipment costs more. However, technical input, calibration and verification of data and professional information technology support are needed to ensure the accurate collection, storage and transmission of data. Due to the lack of experience in use by shipowners, the carbon dioxide emission monitoring system is still out of reach.

4.2   Applicability Analysis of Carbon Emission Calculation Method of Ship

4.2.1   The Calculated Values of IMO Factors are Different

According to the calculation formula of EEOI, the ship energy efficiency operation index is mainly affected by the actual consumption of fuel oil and the carbon dioxide emission factor, which is related to the fuel type, and the fuel consumption is mainly affected by the diesel engine type, ship speed, cargo capacity and other factors. Therefore, it is not a simple linear correlation between the EEOI value and the influence factors.

4.2.2   The Monitoring Calculation of Flow Meter Makes the Statistical Result Close to the Actual Value

While monitoring all kinds of carbon emission sources, the flow meter monitoring method can take into account the residual fuel in the fuel pipe system, so the calculated fuel consumption of ships is close to the actual value. However, in the practical application process, the influence of environmental factors such as wind, wave and flow should also be considered, as well as the ship's own factors such as ship's dirty bottom (which can be multiplied by a dirty bottom coefficient according to the ship's dock repair period), the aging degree of the machine, mechanical transmission efficiency and acceleration distribution map [10].

4.2.3   Monitoring of Field Exhaust Gas Flow Meter Makes the Statistical Results More Accurate

On-site monitoring of carbon emissions, accurate access to fuel combustion carbon emissions. Mass ratio method depends on the accuracy of fuel measurement data, and its practical application value is relatively small. The quantitative method of time ratio method is better than the mass method, provided that the boiler combustion condition is consistent with the monitoring period. The load method is reliable because it considers the influence of boiler.

5   Optimization Measures of Ship Carbon Emission Statistics

5.1   Optimize Monitoring Methods and Explore New Perspectives on Emission Reduction

In terms of improving the efficiency of ship operations and the potential to adapt to future policies, supply single tracking, liquid level measurement only provides total fuel consumption, and direct carbon emission measurement only provides total carbon emissions, while flow meter monitoring provides real-time feedback on the amount of all exhaust gases emitted by the ship in each range. In terms of promoting emission reduction, single-supply tracking and liquid level measurement monitoring have been widely used, which cannot provide a new perspective for emission reduction. However, flow meter monitoring and direct carbon emission measurement methods are not widely used, which have great potential for emission reduction. Therefore, we should give full play to the monitoring methods under digital mode.

5.2   The Correlation Degree of Impact Factors is Analyzed to Improve the Quantification Accuracy of Carbon Emissions

Different quantitative methods are affected by different factors. The paper analyzes the three main influencing factors of EEOI fuel consumption, cargo carrying capacity and voyage distance. Under the condition that two of the factors remain unchanged, the vessel energy efficiency operating index shows a linear positive correlation, linear negative correlation and nonlinear negative correlation with fuel consumption, cargo carrying capacity and voyage distance respectively. Therefore, it is necessary to improve the accuracy of carbon emission statistics by analyzing the correlation degree of each impact factor through real ship data.

6   Conclusions

Although the MRV rules allow applicable ships to use any of the four types of carbon emission monitoring methods described above, ship borne carbon dioxide gas monitoring equipment is not widely used. Future carbon emission calculation is mainly based on fuel consumption monitoring, namely fuel supply single tracking, fuel tank level monitoring and flow meter monitoring methods. According to the quantification method of ship carbon emission, it is found that there is a non-simple linear relationship in the statistical process of ship carbon emission which is affected by various factors.

Acknowledgement. Zhejiang University Students' Science and Technology Innovation Activity Plan and New Miao Talent Project (2021R411041); Zhejiang Ocean University's National innovation and entrepreneurship training program for College Students (202010340035).

References

1. Yihong, W., Fuqiang, L.: Exploring low-carbon operation and practicing green development. China Ports **08**, 25–28 (2018)
2. Hailin, Z., Shuaijun, W., Hu, L.: Carbon emission analysis of ships based on AIS data. China water Trans. (Second Half) **14**(03), 157–158 (2014)
3. Peng, W.: Research on economic speed in ship operation. Dalian Maritime University (2016)
4. Zhenyang, Z.: Research on monitoring means and limitation methods of ship carbon emission. Dalian Maritime University (2015)
5. Shanshan, Y., Chuanxu, W.: Ship speed optimization under different carbon emission control policies. J. Dalian Marit. Univ. **41**(03), 45–50 (2015)
6. Qingbin, C.H.E.N.: Discussion on EEDI technical emission reduction measures. J. Jimei Univ. (Nat. Sci.) **17**(6), 454–459 (2012)
7. Junkai, N.I.: Research on ship energy efficiency operational indicator. Shanghai Jiao Tong University, Shanghai (2010)
8. Bin, L.I.: Introduction and analysis of ship energy efficiency design index and energy efficiency operation index. World Shipping **35**(3), 23–26 (2012)
9. Qiang, W.A.N.G.: Effects and countermeasures of executing new ship energy efficiency design index (EEDI). Guangdongship Build. **4**, 21–23 (2015)
10. Qiang, W.A.N.G.: Ship calculation and analysis of ship energy efficiency operation index. Navig. Technol. **6**, 26–28 (2015)

Research on Vibration Phase Measurement Technology Based on Laser Interference Waveform Demodulation

Hong Cheng[1] and Sha Zhu[2(⊠)]

[1] Zunyi Institute of Products Quality Inspection and Testing, Zunyi 563000, China
[2] National Institute of Measurement and Testing Technology, Chengdu 610021, China
10949999@qq.com

Abstract. Laser interference frequency ratio counting method is widely used in absolute vibration standard devices. However, it can only measure amplitude, and cannot get vibration phase. This paper studies a method of measuring vibration phase by using the relationship between laser interference fringes and vibration displacement of vibrostand, and gives a mathematical model of vibration phase measurement by laser interference waveform demodulation. Based on the homodyne Michelson laser interferometer, this method uses a high-speed data collector to synchronously collect laser interference signals and sensor signals, and demodulates the vibration waveform on the virtual instrument software platform, realizing the real-time measurement of vibration amplitude and phase.

Keywords: Vibration metrology · Waveform demodulation · Homodyne Michelson interferometer · Phase frequency characteristic

1 Introduction

Homodyne Michelson interferometer is a type of laser interferometer widely used in the field of vibration metrology. At present, the fringe counting method established by the homodyne Michelson interferometer is used as the absolute measurement method of vibration. However, it can only measure the vibration amplitude, and cannot measure the vibration phase [1]. The absolute measurement method of vibration amplitude and phase studied in this paper is realized on the basis of homodyne Michelson laser interferometer, using high-speed data acquisition card and virtual instrument technology. While the vibrostand produces a displacement of a half laser wavelength, an interference fringe will be produced. The interference fringe will output a sinusoidal signal similar to frequency modulation through a photomultiplier tube. The data acquisition system collects this signal with the source signal as the gate signal, and then demodulates the vibration displacement waveform on the virtual instrument software platform, realizing the simultaneous measurement of the amplitude and phase of the standard vibrostand and the vibration sensor data [2]. This method realizes the automatic measurement of vibration amplitude and phase, provides a new means for the phase calibration of vibration sensors.

2 Vibration Parameters

In sine linear vibration measurement, there are three basic vibration parameters: displacement, velocity and acceleration, of which acceleration is the main parameter. In the field of engineering applications, vibration accelerometer is generally used to obtain vibration acceleration values. For the verification and calibration of vibration accelerometers, the complex sensitivity of vibration accelerometer sets is obtained by comparison method and absolute method [3, 4].

The complex sensitivity of accelerometer is expressed by the complex-valued function of frequency:

$$\overline{S}_a = \hat{S}_a \exp j(\varphi_u - \varphi_a) \tag{1}$$

wherein, $\hat{S}_a = \frac{\hat{u}}{\hat{a}}$ is the sensitivity amplitude of the accelerometer; $\hat{u}$ is the amplitude of the output voltage u of the accelerometer; $\hat{a}$ is the amplitude of the speed a; φ_u is the initial phase of the output signal of the accelerometer; φ_a is the initial phase of the vibrostand acceleration; $\Delta\varphi = \varphi_u - \varphi_a$ is the phase shift of the complex sensitivity of the accelerometer. The phase shift in formula (1) is the difference between the electrical output phase of the accelerometer and the mechanical vibration input phase. The variation curves of accelerometer sensitivity amplitude $\hat{S}_a$ and phase shift $\Delta\varphi$ in frequency domain are called amplitude-frequency characteristics and phase-frequency characteristics of sensitivity respectively [5].

3 Homodyne Michelson Laser Interferometer

Homodyne Michelson laser interferometer is used in this paper. The measurement principle is shown in Fig. 1. The emitted laser beam is divided into two beams through the beam splitter: the reflected beam (reference beam) is reflected by the reference mirror, then returns to the beam splitter, and is emitted to the photodetector through the beam splitter; the transmitted beam (measuring beam) is reflected by the mirror fixed on the acceleration, then returns to the beam splitter for reflection, and is also emitted to the photodetector [6]. Let the light intensity of the reference beam be E1 and the light intensity of the measuring beam be E2.

$$E_1 = E_{10} \cos(\omega_0 t - \frac{2\pi L_1}{\lambda} + \phi)$$
$$E_2 = E_{20} \cos(\omega_0 t - \frac{2\pi L_2}{\lambda} + \phi) \tag{2}$$

where in: E10, E20—light intensity amplitudes of the reference beam and the measuring beam; L1, L2—optical paths of the reference beam and the measuring beam;
$\quad \omega 0$—optical angular frequency;
$\quad \lambda$—light wavelength;
$\quad \phi$—initial phase of light wave.

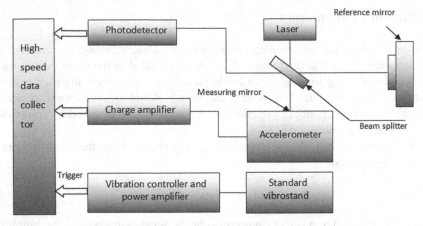

Fig. 1. Schematic diagram of measurement principle.

The photodetector responds to the average intensity of the optical signal, and the photocurrent I

$$I \propto (E1 + E2)2 \tag{3}$$

Substitute the formula (2) into the formula (3) and discard the DC part that the photodetector cannot respond to. The AC voltage U of the photoelectric signal taken out from the sampling resistor is:

$$U = kE_{10}E_{20} \cos[\frac{2\pi}{\lambda}(L_2 - L_1)] \tag{4}$$

$$L_1 - L_2 = d_0 cos(\omega t + \phi_0) + L_0$$

wherein: d0cos(ω t + ϕo) is the vibration displacement; d0 is the single peak of vibration displacement; ω is the angular frequency, and ϕ_o is the initial phase;

L0—optical path difference without vibration, which is a constant phase after simplification. If there is motion interference, it can be expressed as φ0(t).

The formula (4) can be written as

$$U = U_0 \cos[\frac{4\pi d_0 \cos(\omega t + \phi) + L_0}{\lambda}] \tag{5}$$
$$= U_0 \cos[\varphi(t) + \varphi_0(t)]$$

wherein:

$$\varphi(t) = \frac{4\pi d_0}{\lambda} \cos(\omega t + \phi); U_0 = kE_{10}E_{20}$$

It can be seen from the equation that the interference waveform of sinusoidal vibration (that is, the AC voltage U of the photoelectric signal taken out by the sampling resistor) is a signal similar to the frequency modulation waveform, as shown in Fig. 2.

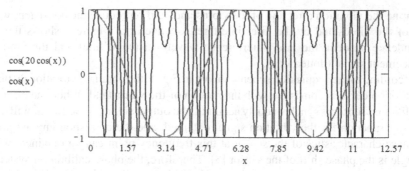

Fig. 2. Correspondence between photoelectric signal (solid line) and vibration displacement signal (dotted line).

4 Laser Interference Waveform Demodulation

The sinusoidal interference waveform is a signal similar to the frequency modulation waveform. If high-speed data acquisition is used to collect a series of zero-crossing time ti (I = 1, 2, 3, 4,...) of photoelectric signal from positive direction (or negative direction) (as shown in Fig. 2). Note that t1, t2... are unequal interval time.

The displacement increment between two adjacent points in the sequence is $\pm\lambda/2$. If the homodyne Michelson interferometer is used, the Z of zero-crossing points from the positive (or negative) direction in N whole vibration periods is calculated, which is equal to the number in the fringe counting method. Based on this, the vibration displacement d0 = λZ/8N can be obtained. Its technical index is the same as that of the fringe counting method [7].

For the vibration displacement signal x(t),

$$x(t) = d0\sin(\omega t + \phi) \tag{6}$$

$$\Delta x = \frac{\lambda}{2} = (dx/dt)(t_{i+1} - t_i) = \omega d_0(t_{i+1} - t_i)\cos(\omega t_i + \varphi)$$
$$= 2\pi f d_0 \cos(\omega t_i + \varphi)/\Delta f(t_i) \tag{7}$$

Displacement difference Δx between two sampling points (7)
Wherein: $\Delta f(t_i) = 1/(t_i + 1\text{-}t_i)$.
From this

$$x'(t_i) = d_0\cos(\omega t_i + \phi) = \frac{\lambda}{4\pi} \times \frac{\Delta f(t_i)}{f} \tag{8}$$

x' is a new vibration displacement signal with the same amplitude and angular frequency as the original vibration displacement signal x(t), it has a phase shift of 900.

Although the homodyne Michelson interferometer cannot identify the vibration direction, when cooperating with the sinusoidal signal driving the vibrostand and referring to the direction of the zero-crossing point of the signal source, we can completely determine the vibration direction of the vibrostand, and reproduce the consistent wave-form of the vibrostand by the laser interference waveform. Figure 2 shows the cor-respondence between the laser interference waveform (solid line) and the vibration displacement signal (dotted line).

According to the acquisition scheme shown in Fig. 1, the obtained vibration displace-ment motion curve is consistent with the motion of the vibrostand. When the output of the calibrated sensor is synchronously acquired, the output signal of the sensor will make a transfer function for the vibration signal obtained, and the corresponding amplitude and phase characteristics of the sensor at this frequency point can be obtained, where the angle is the phase shift of the sensor [8]. Therefore, the phase calibration system of vibration measurement system is constructed. To get better results, the transformed x' can be approximated by a trigonometric function. Namely:

$$x'(t_i) = A \cos \omega t_i + B \sin \omega t_i + C \tag{9}$$

Similarly, A, B and C can be obtained by the least square method, and the vibration displacement and phase can be calculated.

5   Experiments and Results

When the vibrostand is displaced a half of the laser wavelength, an interference fringe will be produced. The interference fringe outputs a sinusoidal signal similar to frequency modulation through a photomultiplier tube, and the pulse signal and the standard signal source signal are input into the data acquisition system at the same time. The data acquisition system counts the pulse signal with the source signal as the gate signal [9]. Then, the pulse signal is converted into a sequence of pulse numbers, and the number of pulses corresponds to the displacement value in the cycle time. The software platform of the measurement system is based on the signal frequency solved by the time interval of peak point. By searching the maximum value of pulse sequence and correcting the three points before and after the maximum value by proportional operation, the maximum point and its position are obtained, so as to determine the peak/bottom position of waveform. Then the signal amplitude is solved by the number of pulse sequences at the peak. The pulse sequence is used to fit and demodulate the reproduced displacement waveform and the amplitude and phase of acceleration [10].

This method can realize the calculation and display of waveforms and parameters such as laser interference signal fitting waveform, signal source input waveform, demod-ulated displacement waveform, sensor input waveform, peak acceleration value, single peak displacement value, distortion degree and phase shift value. The experimental data compared with those of the sine approximation method measurement are shown in Table 1, and the sensor used is quartz flexible acceleration sensor. From the data com-parison, it shows that the phase measured by this method is basically consistent with that measured by sine approximation method, which proves that this method is accurate and feasible.

Table 1. Comparison between sine approximation method and vibration phase measurement method based on laser interference waveform demodulation.

Comparison set	Set A phase (°)														
Frequency	0.103	0.206	0.5078	1.0156	2.0313	3.0469	5	8.125	10	16.25	20	40	60	80	100
Phase shift (sine approximation method)	−4.3	−0.7	−1.7	−3.5	−0.1	−0.1	−0.2	−0.4	−0.4	−0.7	−0.9	−19	−2.8	−4.2	−5.7
Phase shift (this method)	−4.7	−1.2	−1.8	−3.6	−0.1	−0.2	−0.2	−0.4	−0.4	−0.8	−1.1	−1.8	−2.9	−4.2	−5.6
Comparison set	Set B phase (°)														
Phase shift (sine approximation method)	−4.3	−0.7	−1.6	−3.3	−0.0	−0.1	−0.1	−0.2	−0.3	−0.4	−0.6	−0.8	−1.8	−2.5	−3.9
Phase shift (this method)	−4.7	−1.2	−1.8	−3.5	−0.1	−0.2	−0.2	−0.4	−0.4	−0.8	−0.8	−1	−1.8	−2.8	−4

6 Conclusion

The method of measuring vibration phase based on homodyne Michelson laser interferometer is a new method of absolute phase measurement besides sine approximation method. Compared with the sine approximation method, the measurement system realized by this method has simpler structure, lower establishment cost, automatic control, automatic measurement, real-time display and other functions, It's also more accurate for the parameter control of the vibrostand.

References

1. ISO 16063-11: Methods for the calibration of vibration and shock transducers. Part 11: Primary vibration calibration by laser interferometry
2. Wabinski, W., von Martens, H.J.: Time interval analysis of interferometer signals for measuring amplitude and phase of vibrations. Physikalisch-Technische Bundesanstalt (PTB) Fürstenwalder Damm 388,12587 Berlin, Germany
3. JJF 1265-1990: Technical specification for low frequency vertical vibration reference operation
4. JJF 1263-1990: Technical specification for low frequency horizontal vibration reference operation
5. National Metrological Verification Regulation. JJG233-2006. Piezoelectric Accelerometer
6. Yang, Z., Dacheng, L., et al.: Realization method and characteristics of heterodyne laser interferometer. Metrol. Technol. **7**, 2–5 (1996)
7. Jiamin, X., Wei, Z., et al.: Vibration compensation method based on coefficient search for correcting interference fringes of atomic gravimeter. J. Phys. (2022)
8. Mingjian, M.: Data Acquisition and Processing Technology, 2nd Edition. Xi'an Jiaotong University Press, Xi'an (2005)
9. Leping, Y.: LabVIEW Programming and Application. Electronic Industry Press, Beijing (2001)
10. Xihui, C., Yinhong, Z.: LabVIEW8.20 Programming from Beginner to Proficient, 1st Edition. Tsinghua University Publishing House, Beijing (2007)

An Accurate Algorithm for Calibration-Free Wavelength Modulation Spectroscopy Based on Even-Order Harmonics

Yihong Wang, Bin Zhou(✉), Bubin Wang, and Rong Zhao

School of Energy and Environment, Southeast University, Nanjing 210096, China
zhoubinde@seu.edu.cn

Abstract. In our recent work we have shown that even-order harmonics algorithm is a promising measurement method for wavelength modulation spectroscopy techniques. However, the earlier even-order harmonics algorithm was based on an approximate description of the Voigt function given by a pseudo-Voigt approximation model. Therefore, the accuracy of the measured gas parameters cannot be better than the accuracy limit of this approximation. In order to improve the measurement accuracy of earlier algorithm, this paper proposes an updated version based on much more accurate Voigt approximation to implement calibration-free wavelength modulation spectroscopy. The proposed method was validated by a condition-controlled three-zone tube furnace temperature measurement experiment, which showed that the relative error in the calculated gas temperature in the experiment was less than 2.4%.

Keywords: Accurate voigt approximation model · Wavelength modulation spectroscopy · Calibration-free · Even-order harmonics

1 Introduction

Wavelength modulation spectroscopy (WMS) is a mode of tunable diode laser absorption spectroscopy (TDLAS), based on harmonic detection, which has been widely used for the measurement of gas properties, given its advantages of high sensitivity, rapidity, contactless and accuracy [1, 2]. In order to broaden the scope of application of WMS techniques, many calibration-free measurement methods have been proposed, including 1st-harmonic normalization [3–6], multi-parameter harmonic fitting [7] and absorbance lineshape recovery [8–10], etc. It is clear from the considerable literature investigation that the calibration-free $nf/1f$ strategy [4–6], denoted as CF-$nf/1f$, that belongs to the 1st-harmonic normalization method, is being most widespread owing to its simplicity of operation, lack of need for complex analytical models, and universal applicability to a variety of measurement environments, such as high temperature/high pressure [11]. In implementing the CF-$nf/1f$ strategy, the timing-dependent measurement or the simulation of the incident laser intensity, $I_0^S(t)$ in Fig. 1, and timing-dependent wavelength modulation frequency response (WMFR), $v(t)$, need to be determined at an earlier time.

© The Author(s), under exclusive license to Springer Nature Switzerland AG 2022
F. Neri et al. (Eds.): CCCE 2022, CCIS 1630, pp. 155–162, 2022.
https://doi.org/10.1007/978-3-031-17422-3_15

For some given gas parameters, such as gas concentration X, gas temperature T, transition line-centerd wavenumber v_0, integrated absorbance area A, Doppler broadening Δv_D, pressure broadening Δv_C, etc., the simulated transmitted laser intensity, $I_t^S(t)$, can be consequently yielded in accordance with the Beer-Lambert law. One can extrapolate the gas or spectral characteristics when the sum of squared errors (SSE) or the magnitude deviation from the measurement and the simulated WMS signal is minimized once the measured and the simulated transmitted laser intensity, $I_t^M(t)$, and $I_t^S(t)$, are available on-site. Figure 1 describes in detail the implementation route of the CF-$nf/1f$ strategy and the generalized approach to gas parameter retrieval.

Since the harmonic signal is affected by many contributing conditions, such as changes in the nonlinearity of the laser behavior and fluctuations in the laser temperature, an accurately extracting the simulated $nf/1f$, denoted as $^Snf/1f$, has remained a challenge, however [11, 12]. In order to solve the problems mentioned above, we recently developed a novel and rapid algorithm, denoted as *Algorithm I*, for calibration-free wavelength modulation spectroscopy based on even-order harmonics [13]. We have shown that even-order harmonics algorithm is a promising measurement method for wavelength modulation spectroscopy techniques. However, the earlier even-order harmonics algorithm was based on an approximate description of the Voigt function given by a pseudo-Voigt approximation model proposed by Liu et al. [14]. Therefore, the accuracy of the measured gas parameters cannot be better than the accuracy limit of this approximation. In order to improve the measurement accuracy of earlier algorithm, this paper proposes an updated version, denoted as *Algorithm II*, based on much more accurate Voigt approximation to implement calibration-free wavelength modulation spectroscopy.

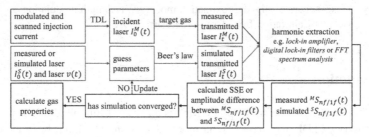

Fig. 1. The implementation route of the CF-$nf/1f$ strategy.

2   Methodology

2.1   Theoretical Background

The detailed theory of *Algorithm I* can be found in reference [13]. In this subsection, only some main conclusions are listed here. It is worth noting that the mathematical symbols in this paper are consistent with reference [13]. The spectroscopic absorbance $\alpha(v)$ can be written as:

$$\alpha(v) = -\ln(I_t/I_0) = A\varphi(v), \tag{1}$$

where v is the instantaneous laser frequency, I_t and I_0 are the transmitted and incident laser intensities, A is the integral absorption area, $\varphi(v)$ is the line-shape function. In *Algorithm I*, the spectroscopic absorbance was approximated by:

$$\varphi(v) = \frac{2}{\pi\lambda}[c_L L(\Delta, m) + c_G \sqrt{\pi \ln 2} G(\Delta, m)], \tag{2}$$

where λ represents the full width at half maximum (FWHM) of the absorption line, $m = 2a/\lambda$ represents the modulation index, a is the modulation depth, $\Delta = 2(v - v_0)/\lambda$ is introduced to describe the relative offset of the laser center frequency v_0, $L(\Delta, m)$ and $G(\Delta, m)$ are the time-dependent peak normalized Lorentzian and Gaussian lineshape functions respectively, c_L and c_G are the weights of the Lorentzian and Gaussian broadening coefficients. In this work, a much more accurate model for the line-shape function described by Voigt function is used for harmonic calculation [15]:

$$\varphi(v) = \frac{2}{\lambda_G}\sqrt{\frac{\ln 2}{\pi}} K(x, y) \tag{3}$$

where λ_G is the Gaussian broadening, $K(x,y)$ is the Voigt function defined as:

$$K(x, y) = \frac{y}{\pi} \int_{-\infty}^{+\infty} \frac{e^{-t^2}}{(x - t)^2 + y^2} dt \tag{4}$$

where x and y are.

$$x = \sqrt{\ln 2}\frac{v - v_0}{\lambda_G} \tag{5}$$

$$y = \sqrt{\ln 2}\frac{\lambda_L}{\lambda_G} \tag{6}$$

where λ_L is the Lorentzian broadening. In the WMS technique, parameter x can be reformulated as:

$$x = \sqrt{\ln 2}\frac{a \cos(\omega t)}{\lambda_G} = 2\sqrt{\ln 2}\frac{\lambda}{\lambda_G} m \cos(\omega t) \tag{7}$$

where ω is the angular frequency of sinusoidal modulation, t means time. In order to simplify the analysis, a lineshape parameter d is introduced as:

$$d = \frac{\lambda_L - \lambda_G}{\lambda_L + \lambda_G} \tag{8}$$

Considering that λ is determined by λL and λG [16], both parameter y and parameter λ/λ_G can be represented by parameter d:

$$y = \sqrt{\ln 2}\frac{1 + d}{1 - d} \tag{9}$$

$$\frac{\lambda}{\lambda_G} = f(d) \tag{10}$$

where $f(.)$ is an anonymous function. Thus the exact mathematical model of the spectroscopic absorbance $\alpha(v)$ represented by parameters m and d can be expressed as:

$$\alpha(v) = \frac{A}{a}F(m,d) \tag{11}$$

where $F(m,d)$ is defined as:

$$F(m,d) = mf(d)\sqrt{\frac{\ln 2}{\pi}}K[2\sqrt{\ln 2}mf(d)\cos(\omega t), \sqrt{\ln 2}\frac{1+d}{1-d}] \tag{12}$$

Obviously, the harmonics of $\alpha(v)$ and $F(m,d)$ are proportional. The amplitude of the n-th harmonic of $F(m,d)$ can be expressed as:

$$H_n = \frac{\omega}{\pi}\int_{-\frac{\omega}{\pi}}^{\frac{\omega}{\pi}}F(m,d)\cos(n\omega t)dt \tag{13}$$

Unfortunately, since there is no analytically exact expression to describe the Voigt function, the n-th harmonic amplitude H_n also have no analytically exact expression.

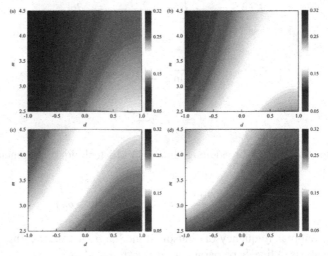

Fig. 2. $H_n(m,d)$ databases for the (a) 2nd-, (b) 4th-, (c) 6th- and (d) 8th- harmonic, respectively.

We note that high precision value of H_n for given m and d can be calculated by numerical method. These amplitude databases $H_n(m,d)$ for the 2nd-, 4th-, 6th- and 8th- harmonic are show in Fig. 2(a), 2(b), 2(c) and 2(d), respectively. The resolution of parameters m and d both are 0.01 for these established databases. Therefore, the accurate estimation of H_n under arbitrary m, d can be derived base on spline interpolation method. Numerical results show that the relative deviation of the estimated H_n is less than 2.37×10^{-6} when $-1 < d \leq 1$ and $2.5 \leq m \leq 4.5$. It is worth noting that the ranges of m and d in this paper are selected based on the optimal parameters of *Algorithm I*. For other parameter ranges or higher precision requirements, these databases can be established in the same way. In addition, the raw data with double precision in Fig. 2 can be found at: https://www.mathworks.com/matlabcentral/fileexchange/101554.

2.2 Fundamentals of the Proposed *Algorithm II*

Figure 2 show that the amplitude of H_n decrease as the even harmonic order n increases. Consequently, the first several nonzero even harmonics should be adopted to achieve higher signal-to-noise ratios (SNRs) of the WMS measurement. In *Algorithm II*, the parameters m and d can be calculated by solving the following minimization problem:

$$(m, d) = \arg\min_{m,d}\{[\frac{h_4}{h_2} - \frac{H_4(m, d)}{H_2(m, d)}]^2 + p_1[\frac{h_6}{h_2} - \frac{H_6(m, d)}{H_2(m, d)}]^2 + \dots$$
$$+ p_k[\frac{h_{2k+4}}{h_2} - \frac{H_{2k+4}(m, d)}{H_2(m, d)}]^2\}$$
(14)

where h_2, h_4, h_6, …, are the measured harmonics amplitudes, p_1, p_2, …,p_k are the corresponding weight coefficient. In this work, only the first three nonzero even harmonics are adopted and the corresponding weight coefficient are both set to 1. Therefore, Eq. 14 can be reformulated as:

$$(m, d) = \arg\min_{m,d}\{[\frac{h_4}{h_2} - \frac{H_4(m, d)}{H_2(m, d)}]^2 + [\frac{h_6}{h_2} - \frac{H_6(m, d)}{H_2(m, d)}]^2\}$$
(15)

With the parameters m and d in hand, the FWHM of the Voigt line-shape can be calculated by.

$$\lambda = 2a/m$$
(16)

and thus the integrated absorbance area A can be calculated by.

$$A = \frac{ah_2}{H_2(m, d)}$$
(17)

The integrated absorbance areas obtained from two individual transitions can be used to infer the gas temperature using two-line thermometry [17, 18]. With the temperature in hand, the gas concentration can be calculated from one of the integrated absorbance areas.

3 Experimental Verification

For a demonstrative purpose, we use the presented *Algorithm II* to address the same experimental data that has been published in our recent work [13]. It should be noted that the parameters employed for signal processing in this paper, such as the low-pass cutoff frequency, are consistent with the previous ones. The experimental measurement system is established based on the schematic diagram shown in Fig. 3.

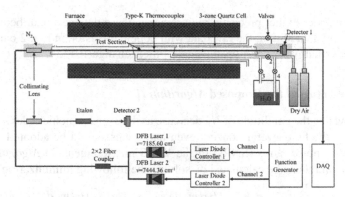

Fig. 3. Experimental setup.

Since the details of the experiments setup can be find in reference [13], we directly give the comparison of temperature measurement results with different algorithms here. Over the scope of 773–1273 K, the temperature values measured by TDLAS system were evaluated in comparison with those measured by thermocouples (Type-K) with an incremental temperature of 100 K. Figure 4(a) illustrates that the temperature measured by TDLAS system with these two algorithms are in good agreement with the temperature measured by thermocouples. The maximum relative errors of temperature measurements for *Algorithm I* and *Algorithm II* are 2.59% and 2.40%, respectively, as shown in Fig. 4(b). Experimental results show the effectiveness of the proposed algorithm.

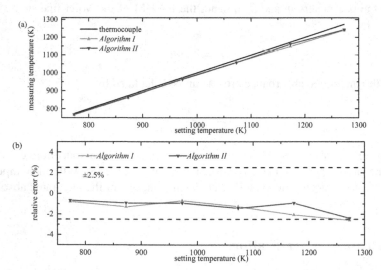

Fig. 4. The temperature measurement results of the two algorithms are compared. (a) the absolute measured temperature, (b) the relative error.

4 Summary

To achieve calibration-free WMS, an accurate algorithm is proposed for accurate retrieval of gas properties based on the even-order harmonics. This algorithm, which is based on much more accurate Voigt approximation, is an updated version of *Algorithm I* proposed in our recent work. The proposed *Algorithm II* was validated by a condition-controlled three-zone tube furnace temperature measurement experiment, which showed that the relative error in the calculated gas temperature in the experiment was less than 2.4%. This work suggests that even-order harmonics algorithm is a promising measurement method for wavelength modulation spectroscopy techniques.

References

1. Liu, C., Xu, L.: Laser absorption spectroscopy for combustion diagnosis in reactive flows: a review. Appl. Spectrosc. Rev. **54**, 1–44 (2019)
2. Wang, Y., Zhou, B., Liu, C.: Sensitivity and accuracy enhanced wavelength modulation spectroscopy based on PSD analysis. IEEE Photonics Technol. Lett. **33**(24), 1487–1490 (2021)
3. Rieker, G.B., Jeffries, J.B., Hanson, R.K.: Calibration-free wavelength-modulation spectroscopy for measurements of gas temperature and concentration in harsh environments. Appl. Opt. **48**, 5546–5560 (2009)
4. Sun, K., Chao, X., Sur, R., Goldenstein, C.S., Jeffries, J.B., Hanson, R.K.: Analysis of calibration-free wavelength-scanned wavelength modulation spectroscopy for practical gas sensing using tunable diode lasers. Meas. Sci. Technol. **24**, 125203 (2013)
5. Goldenstein, C.S., Strand, C.L., Schultz, I.A., Sun, K., Jeffries, J.B., Hanson, R.K.: Fitting of calibration-free scanned-wavelength-modulation spectroscopy spectra for determination of gas properties and absorption lineshapes. Appl. Opt. **53**, 356–367 (2014)
6. Vanderover, J., Wang, W., Oehlschlaeger, M.A.: A carbon monoxide and thermometry sensor based on mid-IR quantum-cascade laser wavelength-modulation absorption spectroscopy. Appl. Phys. B **103**, 959–966 (2011). https://doi.org/10.1007/s00340-011-4570-8
7. Zakrevskyy, Y., Ritschel, T., Dosche, C., Löhmannsröben, H.G.: Quantitative calibration- and reference-free wavelength modulation spectroscopy. Infrared Phys. Technol. **55**, 183–190 (2012)
8. Duffin, K., McGettrick, A.J., Johnstone, W., Stewart, G., Moodie, D.G.: Tunable diode-laser spectroscopy with wavelength modulation: a calibration-free approach to the recovery of absolute gas absorption line shapes. J. Lightwave Technol. **25**, 3114–3125 (2007)
9. Stewart, G., Johnstone, W., Bain, J.R.P., Ruxton, K., Duffin, K.: Recovery of absolute gas absorption line shapes using tunable diode laser spectroscopy with wavelength modulation— Part I: theoretical analysis. J. Lightwave Technol. **29**, 811–821 (2011)
10. Peng, Z., Du, Y., Ding, Y.: Highly sensitive, calibration-free WM-DAS method for recovering absorbance—Part I: theoretical analysis. Sensors **20**, 681 (2020)
11. Du, Y., Peng, Z., Ding, Y.: A high-accurate and universal method to characterize the relative wavelength response (RWR) in wavelength modulation spectroscopy (WMS). Opt. Express **28**, 3482–3494 (2020)
12. Wang, Z., Fu, P., Chao, X.: Laser absorption sensing systems: challenges, modeling, and design optimization. Appl. Sci. **9**, 2723 (2019)
13. Wang, Y., Zhou, B., Liu, C.: Calibration-free wavelength modulation spectroscopy based on even-order harmonics. Opt. Express **29**, 26618–26633 (2021)

14. Liu, Y., Lin, J., Huang, G., Guo, Y., Duan, C.: Simple empirical analytical approximation to the Voigt profile. JOSA B **18**, 666–672 (2001)
15. Mohankumar, N., Sen, S.: On the very accurate evaluation of the Voigt functions. J. Quant. Spectrosc. Radiat. Transf. **224**, 192–196 (2019)
16. Olivero, J.J., Longbothum, R.L.: Empirical fits to Voigt line width: brief review. J. Quant. Spectrosc. Radiat. Transf. **17**, 233–236 (1977)
17. Liu, C., Xu, L., Chen, J., Cao, Z., Lin, Y., Cai, W.: Development of a fan-beam TDLAS-based tomographic sensor for rapid imaging of temperature and gas concentration. Opt. Express **23**, 22494–22511 (2015)
18. Liu, C., Xu, L., Cao, Z., McCann, H.: Reconstruction of axisymmetric temperature and gas concentration distributions by combining fan-beam TDLAS with onion-peeling deconvolution. IEEE Trans. Instrum. Meas. **63**, 3067–3075 (2014)

A Hierarchical Multi-objective Programming Approach to Planning Locations for Macro and Micro Fire Stations

Xinghan Gong[1], Jun Liang[1], Yiping Zeng[1], Fanyu Meng[1,2], Simon Fong[3], and Lili Yang[1(✉)]

[1] Department of Statistics and Data Science, Southern University of Science and Technology, Shenzhen, China
xg2356@columbia.edu, 12032857@mail.sustech.edu.cn, {zengyp, mengfy,yangll}@sustech.edu.cn
[2] Academy for Advanced Interdisciplinary Studies, Southern University of Science and Technology, Shenzhen, China
[3] Department of Computer and Information Science, University of Macau, Macau, China
ccfong@um.edu.mo

Abstract. Fire stations are among the most crucial emergency facilities in urban emergency control system in terms of their quick response to fires and other emergencies. Location planning for fire stations has a significant influence on their effectiveness and capability of emergency responses trading off with the cost of constructions. To obtain efficient and practical siting plans for fire stations, various major requirements including effectiveness maximization, distance constraint and workload limitation are required to be considered in location models. This paper proposes a novel hierarchical optimization approach taking all the major requirements for location planning into consideration and bonds functional connections between different levels of fire stations at the same time. A single-objective and a multi-objective optimization model are established coupled with genetic algorithm (GA) with elitist reservation and Pareto-based multi-objective evolutionary algorithm for model solving. The proposed hierarchical location model is further performed in a case study of Futian District in Shenzhen, and the siting results justify the effectiveness and practicality of our novel approach.

Keywords: Fire station · Hierarchical model · Multi-objective optimization · Genetic algorithm · Multi-objective evolutionary algorithm

1 Introduction

The accelerated process of urbanization poses more and more challenges to the public emergency service in cities than ever before [1]. And there is an urgent need to optimally distribute emergency service facilities, including hospitals and fire stations, in response to the increase of large-scale emergencies [2]. Determination of locations to site emergency service facilities has a significant impact on people's lives and properties [3]. Fire stations,

F. Neri et al. (Eds.): CCCE 2022, CCIS 1630, pp. 163–180, 2022.
https://doi.org/10.1007/978-3-031-17422-3_16

as primary fire rescue providers, are attached with great importance in city planning to ensure public safety and reduce property losses as well [4].

Typical fire station system in China consists of first class and second class fire stations, which are categorized into macro fire stations in this paper. Macro fire stations, as the main executive of fire rescue missions, hold primary responsibility to ensure fire safety in urban areas. However, the construction of macro fire stations consumes a large amount of land and financial resources in terms of setup and operation costs. Besides, macro fire stations are typically assigned with enormous service areas, resulting in a longer rescue time to the edge of service areas. Therefore, expanding only macro fire stations to enhance fire management system is no longer appropriate in current urban planning in regards of financial costs and time based rescue efficiency. The Chinese government proposed a hierarchical fire station structure in "Construction Standard of Urban Fire Stations 2017" ("Construction Standard" in the rest of the texts) [5] to fix this problem. In "Construction Standard", a novel type of fire station, micro fire station, is recommended to mainly serve local communities. Compared with macro fire stations, micro fire stations have a much lower setup and operation cost because of their less area occupancy and smaller service areas. The smaller service areas character further allows micro fire stations to have a much shorter rescue time, bringing high rescue efficiencies to local communities. Therefore, a reasonable location siting of macro fire stations coupled with micro fire stations is of great need for decision makers (DM) to strengthen the urban fire rescue forces.

Related papers show that multiple modeling methods have been conducted in urban fire station location problem, e.g. maximal covering problem (MCP), set covering problem (SCP) and p-median problem (PMP) [6]. Specifically, a fuzzy multi-objective programming combined with GA has been proposed to determine the optimal fire station locations to serve areas with different fire risk categories [7]. Yao *et al.* [8] constructed an bi-objective location model combined with SCP and PMP to locate fire stations adjacent to areas with high fire risks and minimize the total construction number as well. Aktaş [9] *et al.* applied both the SCP and MCP to acquire fire station location in 10 different scenarios in Istanbul. To make the optimal fire station locations more closely fit the actual situation, they employed Istanbul's road networks to determine additional fire station locations. Furthermore, a two ranked level model was developed to minimize response time to fire accidents with allocation strategies fully considered [10]. And Hendrick *et al.* also proposed that the influence of existing fire stations should not be neglected while considering where new ones are placed [11, 12].

Though differences exist in the location modeling techniques for fire station planning, some common principles and constraints are followed in these researches:

- Minimization of financial cost of fire stations. Minimizing the number of fire stations is often needed to meet the budget constraint or reduce the setup and operation costs [13, 14].
- Minimization of response time. Arrival assurance of fire engines at fire scenes within a certain time is the most basic standard the fire service authorities must follow [3, 7, 10].
- Maximization of total service areas. It is desired to cover as many demand points with high risks as possible [9].

- Minimization of the distance between fire stations and high-risk areas. With fire risks varying in space, optimal locations of fire stations should be close to high-risk regions to shorten the response time [15, 16].

Despite of the fact that hierarchical modeling method has been widely applied in emergency medical service systems and telecommunications networks problems [17–19], the existing researches in fire station planning mainly focused on modeling single leveled fire station system and few studies paid much attention to the current hierarchical structure of macro and micro fire stations. Yu *et al.* [3] adopted MCP and SCP models to arrange the locations of macro and micro fire stations with GA applied into solving the hierarchical model. A basic assumption lay in [3] is to relocate all macro fire stations instead of adding new ones. However, in practical application, it is unrealistic and of high costs to site macro fire stations in this way. Besides, insufficient factors have been considered in [3], as two single objective optimization problems with limited constraints are proposed for macro and micro fire station locations. Therefore, this paper aims to construct a more realistic and practical hierarchical location model for macro and micro fire stations based on the government's requirements in "Construction Standard". The mathematical representations for macro and micro fire stations models are single objective programming and multi-objective programming, respectively. Our hierarchical also involves the effect of existing fire stations and real road network in potential location points for fire stations. A case study in Futian District, Shenzhen indicates that our modeling method has strong portability and is applicable to the emergency management authorities.

The main contributions of this paper include:

- This paper provides a practical construction approach for cities without a complete multi-level fire facility system.
- This paper combines the covering location model with geographic information system (GIS) approach to better simulate the real-world situation. And the visualization makes our modelling procedures more understandable.
- Considering the hierarchical location structure of macro and micro fire stations, we take the following aspects into our model construction:

   - Take factors including transportation and rescue efficiency into account.
   - Apply multi-objective programming in the second-level model.
   - Establish an interaction between fire stations with different levels.

The remainder of this paper is organized as follows. Section 2 presents our hierarchical location model for macro and micro fire stations. Section 3 conducts a case study in Futian District, Shenzhen, China, with GA methods applied to solve the model. The results validate the practicality and efficiency of our solving and modeling approaches. Then, the paper is ended with some conclusions in Sect. 4.

2 Hierarchical Location Model

In this section, a hierarchical model is proposed to solve the location problem of macro and micro fire stations. In specific, an extended MCP with distance constraint between neighboring stations, and a combination model of SCP and PMP with capacity and distance constraints are adopted to select locations for macro and micro fire stations, respectively. The following subsections introduces the two models in detail.

2.1 Model Parameters

The following parameters will be applied into our hierarchical location model.

- I: Set of demand point, $i \in I$;
- J_1: Set of potential macro fire station locations, $j_1 \in J_1$;
- J_2: Set of potential micro fire station locations, $j_2 \in J_2$;
- d_{ij_k} : Distance between i and j_k, $k = 1, 2$;
- $d_{jj'}$: Distance between two adjacent fire stations, $j, j' \in J_k, k = 1, 2$. For any two stations located at $j_a, j_b \in J_k, k = 1, 2$, we call the fire station located at j_a "adjacent" to the another fire station located at j_b if the distance between these two stations is shorter than the distance between the fire station at j_a and any other fire stations [7]. The detailed definition of adjacent fire stations is described as follows:

$$d_{j_a j_b} \leq d_{j j_a}, \ \forall j \in J_k, j \neq j_a, j_b, k = 1, 2. \tag{1}$$

- d^{s_1}, d^{l_1} : Minimal and maximal distance between each pair of adjacent macro fire stations, respectively;
- d^{s_2}, d^{l_2} : Minimal and maximal distance between each micro fire station and its adjacent macro fire station, respectively;
- a_i : Demand value of node i. In this study, demand value is defined as the estimated fire risk value;
- N : Number of macro fire stations to be located;
- S : Workload limitation value of micro fire station. In this paper, the workload is defined as the total sum of fire risks covered by each fire station;
- $\Omega_{ij_k} = \{j_k | d_{ij_k} \leq R_k, k = 1, 2\}$: Set of potential locations j_k capable of serving demand point i;
- $\eta_{ij_k} = \{i | d_{ij_k} \leq R_k, k = 1, 2\}$: Set of demand points i that can be covered by potential fire station at j_k;
- Φ : Set of existing macro fire station locations;
- R_1, R_2 : Service radius of macro and micro fire stations, respectively;
- $X_{j_k} = \begin{cases} 1, \text{if a potential fire station is sited at } j_k \\ 0, \text{else} \end{cases}$
- $Y_{ij_k} = \begin{cases} 1, \text{if node } i \text{ is suitably covered by station at } j_k \\ 0, \text{else} \end{cases}$

2.2 Macro Fire Station Model

In our macro fire station model, we propose to keep all the existing macro fire stations and build a limited number of new macro stations to cover as many high-risk communities as possible. And the distance constraints between adjacent stations are adopted to obtain a balanced distribution of newly-built macro stations.

Extended MCP for Macro Fire Stations. The proposed model is formulated as follows,

- Objectives:

$$\text{Max} \sum_{i \in I} a_i Y_{ij_1} \tag{2}$$

- Subject to:

$$\sum_{j_1 \in \Omega_{ij_1}} X_{j_1} \geq Y_{ij_1}, \forall i \in I, \tag{3}$$

$$\sum_{j_1 \in J_1} X_{j_1} = N, \tag{4}$$

$$d^{s_1} \leq d_{j_1 j_1'} \leq d^{l_1}, \forall j_1, j_1' \in J_1, \tag{5}$$

$$d^{s_1} \leq d_{j_1 j_1'} \leq d^{l_1}, \forall j_1 \in J_1, j_1' \in \Phi, \tag{6}$$

$$X_{j_1}, Y_{ij_1} = \{0, 1\}, \forall i \in I, j_1 \in J_1. \tag{7}$$

Equation (2) is an objective aiming to maximize the total demand coverage of communities by macro fire stations, thereby encouraging macro fire station to cover as many high -risk communities as possible. Constraint (3) ensures that each chosen community would be covered by at least one macro fire station. Constraint (4) specifies the number of new macro fire stations to locate. Constraint (5) limits the distance between each pair of new adjacent macro fire stations. Furthermore, if a new macro fire station is adjacent to an existing macro fire station, constraint (6) limits the distance between these two fire stations. Finally, constraint (7) imposes binary coding requirement on decision variables X_{j_1} and Y_{ij_1}.

2.3 Micro Fire Station Model

Micro fire stations have a low setup and operation cost and are convenient to relocate because of their assembled building structure. Therefore, even if the initial allocation of micro fire stations is unreasonable, DM still can relocate micro fire stations with low costs.

In current fire management system, micro fire station is taken as a supplementary rescue force for macro fire stations. In specific, since the macro fire station model is to cover as many high-risk communities as possible, there exists some communities with smaller fire risks that cannot be covered by macro fire stations. Therefore, micro fire stations are sited to cover these communities and provide auxiliary support to macro fire stations in high-risk communities as well. Furthermore, service area of each micro fire station is required to intersect with at least one macro fire station's service area. Therefore, based on these siting requirements, we combine SCP and PMP with a novel objective in maximizing the average distance of adjacent micro fire stations to construct our model.

Spatial Location Model for Micro Fire Station. The proposed model is formulated as follows,

- Objectives:

$$\text{Min} \sum_{j_2 \in J_2} X_{j_2} \tag{8}$$

$$\text{Min} \sum_{i \in I} \sum_{j_2 \in \Omega_{ij_2}} a_i d_{ij_2} Y_{ij_2} \tag{9}$$

$$\text{Max} \frac{\sum_{j_2,j_2' \in J_2} d_{j_2 j_2'}}{|J_2|} \tag{10}$$

- Subject to:

$$\sum_{j_2 \in \Omega_{ij_2}} X_{j_2} \geq 1, \forall i \in I, \tag{11}$$

$$Y_{ij_2} \leq X_{j_2}, \forall i \in I, j_2 \in J_2, \tag{12}$$

$$\sum_{i \in \eta_{j_2}} a_i \leq S, \forall i \in I, \tag{13}$$

$$d^{s2} \leq d_{j_1 j_2} \leq d^{l2}, \forall j_1 \in J_1, j_2 \in J_2, \tag{14}$$

$$X_{j_2}, Y_{ij_2} = \{0, 1\}, \forall i \in I, j_2 \in J_2. \tag{15}$$

The objective (8) is to minimize the total number of newly-built micro fire stations. The objective (9) is to minimize the distance between micro fire stations and communities with higher fire risks. And objective (10) is to maximize the average distance of each pair of adjacent micro fire stations. Constraint (11) and (12) ensure that all the communities can be covered by at least one micro fire stations. Constraint (13) limits the workload of each micro fire station where we define the total sum of fire risk values within the service area as the workload value. Constraint (14) limits the distance between each micro station and its adjacent macro fire station. And finally, binary requirements are imposed in constraint (15).

The combination of the objectives and constraints constitute a multi-objective decision optimization model for the micro fire station location problem, shown as

$$\text{Min } [F1, F2, F3], \tag{16}$$

where $F1 = \text{Obj}1$ (8), $F2 = \text{Obj}2$ (9) and $F3 = -\text{Obj}3$ (10).

3 Case Study

The proposed model is implemented and applied in an empirical study evaluating fire service in Futian District, Shenzhen, China. The interest is to construct a new fire station system while considering the connection between different levels of fire stations and the existing fire stations' effects. Our research scenario here is to keep all the existing macro fire stations and determine optimized locations for new macro and micro fire stations. The model is solved by using Python Geatpy library [20], and all the experiments are run on a computer with an Intel Core i7 2.59 GHz CPU and 8.0 GB RAM. We also combine commercial GIS software, ArcGIS (version 10.6) into selecting potential fire station location points and visualizing our results.

3.1 Study Area

The research area, Futian District, is located in Shenzhen, Guangdong Province, China. Shenzhen is one of the mega-cities in southern China and Futian District is the central area of Shenzhen, which takes up, according to "Shenzhen Statistical Yearbook-2020" [21], 16.88% (454.65 billion Yuan) of Shenzhen total GDP (2692.71 billion Yuan) and has a population density of 20769 person per square kilometers. With such a high economic status and density of population, Futian District is undertaking great pressure from fire control management. Therefore, there exists an urgent need for the city's DM to enhance the fire rescue forces in the region.

3.2 Data Preparation

In the case study, we obtain our potential location points based on the road network constructed from the open street map (OSM). And in our calculation of estimated fire risk value of each community, we acquire the historical fire accident location points between 2014 and 2019 in Futian District from the Emergency Management Bureau of Shenzhen Municipality, and the population density value from the open platform of Shenzhen government [21]. In total, there are 1789 historical fire accident points, 4 existing macro fire station and 95 communities in Futian District. And our novel hierarchical location model is utilized to generate siting locations for macro and micro fire stations.

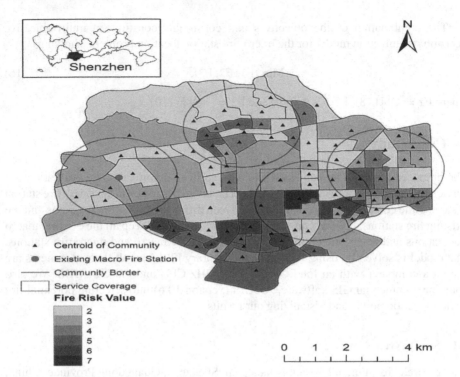

Fig. 1. Study area, existing macro fire stations and fire risk values.

3.3 Location Siting of Macro Fire Station

In "Construction Standard", the government requires that the response time of macro fires station should be within 4 min, and the scale of fire stations' service area is also limited for the service area of first-class fire station should not exceed 7 km^2 and second-class fire station's under 4 km^2. Besides, "Construction Standard" provides the follow equation to calculate service area of fire station

$$A = 2P^2, \qquad (17)$$

where A is the service area of fire station and P is the coverage radius of service area. Thus, taken these regulations and real road structure into account, we propose the following weighted method to get the radius of service area in our model

$$R_1 = \sqrt{\tfrac{1}{2}(\beta \cdot A_1 + (1 - \beta) \cdot A_2)}, \beta \in (0, 1), \qquad (18)$$

where A_1 and A_2 are the recommended service areas of first and second class fire stations respectively based on "Construction Standard". R_1 represents the service radius of macro fire station in the model, and β is the weight value determined by DM. In the case study, to reflect the priority of first class fire station and combined with fire service authorities' opinions, the value of β is set to 0.7 and with $A_1 = 7\,\text{km}^2$ and $A_2 = 4\,\text{km}^2$. Then, the service radius of macro fire station R_1 in the model is set as 1.746 km. In this section, we introduce the key steps to obtain the locations of new macro fire stations as follows:

1. Calculation of Estimated Fire Risk Values: In the hierarchical model, the demand value a_i is defined as the estimated fire risk value of each community. By this way, in the first level of the hierarchical model, newly-built macro fire stations are required to locate near high-risk communities while satisfying all the constraints. Thus, we first need to calculate the estimated fire risk value.

Table 1. The fire risk value of communities in terms of the number of historical fire accidents.

The number of historical fire accidents	The number of communities	Risk value r_a
$2 < n \leq 14$	45	1
$14 < n \leq 26$	33	2
$26 < n \leq 46$	10	3
$46 < n \leq 99$	7	4

Table 2. The fire risk value of communities in terms of population density.

Population density ρ	The number of communities	Risk value r_p
$0.26 < \rho \leq 3.60$	53	1
$3.60 < \rho \leq 8.02$	28	2
$8.02 < \rho \leq 13.37$	12	3
$13.37 < \rho \leq 24.85$	2	4

Regional fire risk in urban communities is closely interrelated with population size and the frequency of fire accidents within local communities. Therefore, in the case study, we combine the population density with the number of fire accidents in each community to calculate the estimated fire risk value. First, we apply natural breaks [22] in our model to rank fire risk levels of communities. Natural breaks has advantage in searching natural breakpoints which have statistical meanings in sequences, and with these obtained breakpoints, the sequence can be classified into groups with similar properties. As presented in Table 1 and 2, we rank the fire risk levels of communities from population density and total number of fire accidents with each aspect dividing into four categories using natural breaks method. Then, we combine them with a weight parameter γ, $\gamma \in (0, 1)$, to get the final estimated fire risk values of communities according to equation

$$a_i = \gamma \cdot r_a + (1 - \gamma) \cdot r_p. \tag{19}$$

For general purposes, we adopt the parameter γ as 0.5 [3]. we then map fire risk values into each corresponding community in Fig. 1, where communities with deeper colors have higher fire risk values. Specifically, Fig. 1 shows the study area, estimated fire risk and the location of existing macro fire stations. In the case study, centroid points are taken from main residential area of each community and we assume that fire stations can serve communities as long as their service areas can cover their centroids. It can be seen from Fig. 1 that existing macro fire stations have a unbalanced distribution for communities with deeper colors in the southern area of Futian District cannot be served by these fire stations. Therefore, there exists an urgent need to construct new macro fire stations in this region.

2. Calculation of the Number of New Macro Fire Stations: Tzeng et al. [16] and Yang et al. [7] calculate the number of new fire stations to be built by balancing the total setup and operating cost of fire stations and total lost cost of incidents in a given area. The detailed expression of the optimal cost model is summarized as follows,

$$\text{Min} f(N) = N \times SC + \alpha \times TLC \times e^{-N}, \tag{20}$$

where SC is the setup and operating cost of each fire station, TLC is the total lost cost of incidents in a given area and N is the number of new fire stations to be built. And by setting the derivative of $f(N)$ as zero, the number of new fire stations N is given as

$$N = \text{int}(\log TLC - \log SC + \beta). \tag{21}$$

However, for the initial lost cost model in [16] with $\alpha = 1$ in (20), the value of SC is much higher than TLC in megacities (e.g. Shenzhen) during normal years based on the data from open data platform of Shenzhen government [21]. As a result, this causes $N < 0$ in (21) and cannot be applied into our model. As for the lost cost model with $\alpha > 1$, the historical record utilized to calculate α is normally hard to obtain because of the loss of relative data in fire management system. Therefore, our model proposes to obtain the number of new macro fire stations to be built based on the coverage area of macro fire station and total area of given research region with

$$N = \lceil \tfrac{TAR - N_e \times SAM}{SAM} \rceil, \tag{22}$$

where TAR is the total area of research region, N_e is the number of existing macro fire stations and SAM is the service area of each macro station. In (22), we round up the value because the equation does not consider in the overlapping of service area in terms of the irregular-shape character of research regions which would lead to the unavailability of the actual number of new macro fire stations. Hence, the actual number of new macro fire stations is higher than the result of (22) and we round up the value to solve this problem.

In the case study, the total area of Futian District is 78.66 km^2 and urban area takes up 75% of it (i.e. 58.995 km^2). Also, with the service area limitation in "Construction Standard" and our model assumptions, the service area of each macro fire station is 9.58 km^2. Then, according to (22), 3 new macro fire stations need to be sited in the model.

3. Selection of potential location points for macro fire stations: The division of communities in Futian District is based on road network, and according to "Construction Standard", locations of fire stations are required to have high transportation accessibility. Therefore, in the case study, we first find all intersections in the road network of Futian

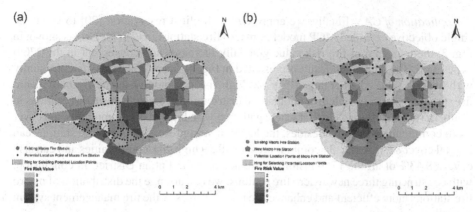

Fig. 2. Final potential location points for macro fire stations (a) and micro fire stations (b).

District and take these intersection points as potential location points for macro fire stations. Specifically, since the number of newly-built macro fire stations has been settled in Sect. 3.3-2, we propose to expand choices of potential location points for macro fire stations by taking additional points every 200 m within each road segment. However, since each road segment is not exact multiple of 200, we delete potential location points within 50 m of road intersections which have higher priorities than other points because of their transportation convenience. Besides, as shown in Fig. 2(a), in order to fulfill the distance limitation between each new macro fire station and its adjacent existing macro fire station, we take final choices of potential location points within the ring centered at location points of existing macro fire stations with a width of $d^{l_1} - d^{s_1}$, where

$$d^{l_1} = 2R_1 + \varepsilon, \tag{23}$$

and d^{s_1} is obtained by requiring the overlapping area of adjacent macro fire stations cannot exceed 30% of the total service area. In (23), each pair of adjacent macro fire stations should not be too far from each other with a tolerance of $\varepsilon = 0.05$ km. and we obtain total 239 potential locations points for new macro fire stations in Fig. 2(a).

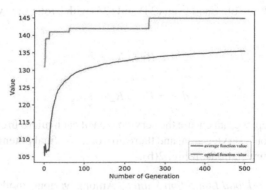

Fig. 3. Total sum of fire risk values of elite individual in each generation.

4. Application of GA: Finally, we apply GA with elitist reservation [20] to solve the single objective extended MCP model of macro fire stations in (2)-(7). And as shown in Fig. 3, the total sum of fire risk value gradually increases and convergent in the 270th generation. The final siting result is presented in Fig. 4(a), and if we classify community with fire risk value equal or larger than 4 as high-risk community, the newly-built macro fire stations solve the problem of unbalanced distribution of existing macro fire stations in southern region of Futian District where gathered with high- risk communities and cover more communities as well. Besides, the location siting result of macro fire stations can in total cover 87.18% (i.e. 34 communities) of all 39 high-risk communities and can also cover 88.43% of all the 1789 historical fire accidents in Futian District. In conclusion, with constructing three new macro fire stations, we can arrange the distribution of macro fire stations more efficient and enhance fire rescue forces of the fire management system as well.

3.4 Location Siting of Micro Fire Stations

According to "Construction Standard", micro fire stations target at serving communities and with a limitation of 3 min response time and service area under 1 km^2. Then, based on (17), service radius of micro fire station is set to 1 km. The demand value a_i of communities is the same with the fire risk value obtained in Sect. 3.3–1. The key steps on obtaining siting results of micro fire stations are introduced as follows:

1. Selection of Potential Location Points: In the case study, the potential location points of micro fire stations are also obtained from the road network in Futian District. The difference is that we only take intersection points of road network this time. This is because the number of micro fire stations to be sited is not fixed and with too many potential location points, convergence process of the GA applied to solve the model will be too time-consuming, leading to a low efficiency to get the siting results. So, similar method in Sect. 3.3–3, we obtain final potential location points within the ring which centers at all the macro fire stations including the existing ones and the results got from the macro fire station model, and with a width of $d^{l2} - d^{s2}$, as presented in Fig. 2(b). The values of d^{l2} and d^{s2} are obtained from (24) and (25) with a tolerance value $\varepsilon = 0.05$ km respectively,

$$d^{l2} = R_1 + R_2 + \varepsilon, \tag{24}$$

$$d^{s2} = R_1 - R_2 - \varepsilon. \tag{25}$$

Through this step, we can ensure the service area of each micro fire station is adjacent to a macro fire station's service area, and there are total 165 final potential location points selected for micro fire stations in Fig. 2(b).

2. Calculation of Workload Limitation Value: Among various models in solving the location problem of fire stations, a usual assumption is that facilities are without workload limitations which means as long as the location of a fire accident is sited within the service of a fire station, it will be suitably resolved by the fire station. This assumption might work with macro fire stations for their high construction standard with more rescue equipment. However, micro fire stations do not possess such high rescue capability and a micro fire station with excessive communities to manage will lead to unbalancedness and inefficiency in fire management system [3]. Therefore, in the second-level of our model, we add a capacity constraint (13) so as to balance the number of high-risk communities can be covered by micro fire stations. And the workload is working as the upper average total fire risk value which can be covered by each micro fire station. The calculation of workload value is shown as follows,

$$S = \frac{\sum_{i=1}^{M} \max_i \{S_{ij}\}}{M}, 1 \leq j \leq |D_i|, i, j \in Z, \qquad (26)$$

where S_{ij} is the j th workload value from i th run. D_i is the siting solution space of i th run and M is the number of runs.

To obtain the value of S_{ij}, we first take away the workload constraint (13) in the model and acquire the siting results preferred by F2. This is because F2 encourages each micro fire station to cover as many high-risk communities as possible and thereby making each micro fire station reach its upper average total fire risk coverage value. Then S_{ij} is the average sum of fire risk values of each siting result preferred by F2.

In the case study, the model without constraint (13) is solved by using NSGA-II. NSGA-II is a highly efficient MOEA and has been widely applied into solving multi-objective optimization problems [23]. NSGA-II involves an elitist-preserving approach to speed up the performance of the algorithm and help prevent the loss of elitist solutions once they are found. Furthermore, a fast non-dominated sorting approach is applied to reduce the computation complexity, and a crowding distance evaluation mechanism is used to preserve the diversification of Pareto-optimal solutions. Detailed framework of NSGA-II can refer to [24].

Encoding scheme in MOEA also has great effect on the performance of the algorithm [25]. In the case study, unlike [3, 23] and [26] which adopt a mesh-based spatial representation strategy, we obtain our potential location points on the road segments. Therefore, we apply the classical binary encoding scheme with "1" representing the locations to be sited with micro fire stations and "0" representing the otherwise in NSGA-II.

Due to the stochastic nature of evolutionary algorithms, it is necessary to perform several parallel runs to evaluate their performance. Therefore, without loss of generality, 10 independent runs (i.e. $M = 10$) are implemented by using fixed parameter values but with different initial populations [23]. In specific, the maximum number of evolutionary generations is 500. The size of the initial population is limited to 300, the crossover probability is 0.9 and the mutation probability is 0.005. The best final Pareto-optimal front among 10 runs is displayed in Fig. 6(a). As presented in Fig. 6(a), the result shows the ability of NSGA-II in finding a well-convergent Pareto-optimal front.

The performance of NSGA-II is evaluated based on hypervolume (HV) and Spacing indicators. The HV indicator is described as the volume of the space in the objective space dominated by the Pareto front approximation and delimited from above by a reference point [27]. It has been utilized to evaluate the convergence and diversity of MOEA and the bigger the HV is, the batter the performance is. In the case study, we take the nadir point [28] in each generation as the reference point and detailed calculation algorithm of HV can refer to [29]. The Spacing indicator captures the variation of the distance between elements of a Pareto front approximation and a lower value of the MOEA is considered to be better [27]. This paper adopts the Spacing metric to evaluate the distribution character of non-dominated solution points along each Pareto front approximation.

First, we analyze the convergence performance of NSGA-II based on HV metric. The values of HV for every generation are shown in Fig. 5(a). We find that the HV values sharply increase for the first 100 generations and finally become stable for 200 generations from 300th to 500th generation, which shows that NSGA-II is able to converge around 300th generation. Furthermore, as seen in Fig. 5(b), non-dominated solution points are able to uniformly distributed along Pareto front approximations from around 100th generation because the spacing values keep stable for 400 generations from 100th generation. It shows the capability of NSGA-II in generating a set of uniformly distributed Pareto optimal solutions. Therefore, it is appropriate to apply NSGA-II into solving our model and the final computation result of workload value S according to (26) is 19.374829.

3. Model Solution and Analysis: The spatial location model for micro fire stations with capacity and distance constraints (8)–(15) is also solved by NSGA-II. Similarly, 10 independent runs are implemented with the same parameters defined in Sect. 3.4–2 and the best result is shown in Fig. 6(b). As presented in Fig. 6(b), it shows the ability of NSGA-II in finding a well-convergent Pareto-optimal front approximation.

The model verification is also based on HV and Spacing metrics. For the evaluation of convergence performance of the algorithm, as shown in Fig. 5(a), NSGA-II converges from around 200th generation for the HV values keep stable for 300 generations from 200th to 500th generation. As for the distribution property of solutions, Fig. 5(b) represents that the Pareto solution sets are able to uniformly distributed starting from 100th generation.

Table 3. Objective values and incident coverage rates of the four siting solutions.

Solution	A	B	C	D
F1	26	86	26	37
F2	187.87	113.98	188.91	140.75
F3	−1.31	−0.42	12	−0.85
Incident Coverage Rate	93.63%	99.27%	94.47%	96.81%

The 275 non-dominated solutions on the final Pareto optimal front of NSGA-II together constitute a candidate pool for the DM, where siting solutions of micro fire stations with different preferences can be chosen as the final location plans. However, there is no best solution among these non-dominated solutions because they cannot dominate each other by definition. In the case study, an equally weighted solution and three extremely preferred solutions are selected among the 275 non-dominated solutions

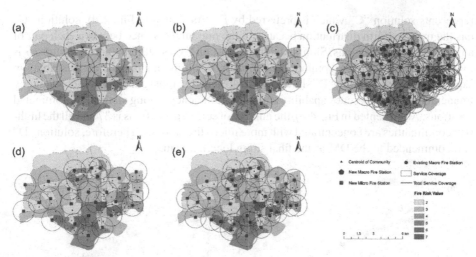

Fig. 4. Siting results of macro fire stations (a) and four representative siting solutions of micro fire stations: A is preferred by F1 (b); B is preferred by F2 (c); C is preferred by F3 (d); and D is the equally weighted solution (e).

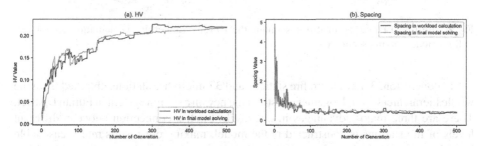

Fig. 5. The values of HV (a) and Spacing (b) in section III-D2 and III-D3.

[26]. The four kinds of solutions are respectively labeled as A, B, C and D. For each solution, statistical analyses are presented in Table 3 and the result visualization is displayed in Fig. 4.

Figure 4(b) represents solution "A" which is preferred by F1. The locations of the 26 micro fire stations have an overall uniform distribution with more concentration around the northern high-risk communities of Futian District. However, not all high-risk communities, especially communities in the southern areas, receive attention from the location solution and each micro fire station located in the southern areas obviously has to cover excessive high-risk communities, which shows a unbalanced distribution of the solution. This is because solution "A" only taking the number of stations to be constructed as the most important factor. Figure 4(c) represents solution "B" which is preferred by F2. The number of selected locations is 86 and the solution achieved the distribution of more concentration around all high-risk communities. However, the high overlap rate around these high-risk communities shows an economic inefficiency of the solution and therefore, the solution "B" is not recommended to the DM. Figure 4(d) represents solution "C" which is preferred by F3. As shown in Fig. 4(d), solution "C" has the most uniform distribution because of the maximal distance between each pair of adjacent micro fire stations. Similar defect in solution "A", southern micro fire stations is overloaded with high-risk communities to cover, resulting in a unbalanced distribution. Equally weighted solution "D" shown in Fig. 4(e) has the most balanced trade-off among financial limitation, rescue capability and rescue efficiency among all the non-dominated solutions. As presented in Fig. 4(e), the number of selected locations is 37 and all the high-risk communities are concentrated with more micro fire stations. Therefore, solution "D" is recommended to the DM as the final siting location plan.

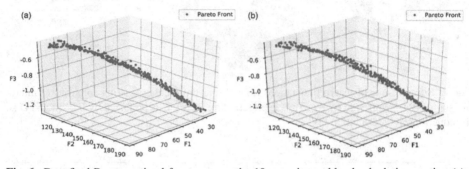

Fig. 6. Best final Pareto optimal fronts among the 10 runs: in workload calculation section (a); in final model solving section (b).

In conclusion, 3 new macro fire stations and 37 micro fire stations obtained from the whole hierarchical model together construct the new fire station system in Futian District. There are three major advantages in our result. First, a connection between different levels of fire stations is constructed in the model, making our results more reasonable in reality. Second, with keeping all the existing macro fire stations and constructing a limited number of new ones, the siting results allows macro fire stations to cover more high-risk communities and micro fire stations, as a supplementary rescue forces in fire management system, cover the remaining ones. Finally, the results fulfill our initial aim to require micro fire stations to reach the fire accident scene at the very beginning while waiting for the adjacent macro fire stations to implement major rescue mission. More specifically, the coverage rate of high-risk communities is 100% and the entire fire station system could also cover 98.66% (i.e.1765) of total historical fire accidents in Futian District as well. Therefore, with our hierarchical location model, a reasonable and efficient location siting instruction is able to be provided to DM.

4 Conclusions

Considering the multi-level structure of fire station system proposed by the Chinese government, this paper constructs a hierarchical model for macro and micro fire stations. In specific, in the first level of the hierarchical model, an extended MCP with distance constraint between adjacent stations is selected for macro fire stations siting because of the fixed number of new macro fire stations to be sited. Then, in the second level of our hierarchical model, a combination model of SCP and PMP with capacity and distance constraints is adopted for micro fire stations, as micro fire stations are established to mainly serve local communities. The hierarchical model is further performed in a case study of Futian District in Shenzhen, and a non-dominated sorting genetic algorithm II (NSGA-II) is proposed to solve the model. NSGA-II is a highly efficient and elitist-preserving multi-objective evolutionary algorithm (MOEA). In the case study, we adopt hypervolume (HV) and Spacing indicators to evaluate the performance of NSGA-II, and the test results show its feasibility in solving the hierarchical model. Furthermore, the final siting results of macro and micro fire stations prove the effectiveness of our model, which can be utilized to generate valid reference for decision makers.

Acknowledgement. This research was supported by National Key R&D Program of China (2019YFC0810700 and 2018YFC-0807000), National Natural Science Foundation of China (71771113, 71704091 and 71804026 No. 72004141) and Shenzhen Science and Technology Plan Project (N0. JSGG20180717170802038) and Basic and Applied Basic Research Foundation of Guangdong Province (No. 2019-A1515111074).

References

1. Farahani, R.Z., Fallah, S., Ruiz, R., Hosseini, S., Asgari, N.: OR models in urban service facility location: a critical review of applications and future developments. Eur. J. Oper. Res. **276**(1), 1–27 (2019)
2. Paul, N., Lunday, B., Nurre, S.: A multi-objective, maximal conditional covering location problem applied to the relocation of hierarchical emergency response facilitie. Omega **66**, 147–158 (2017)
3. Yu, W., Chen, Y., Guan, M.: Hierarchical siting of macro fire station and micro fire station. Environ. Plan. B: Urban Analytics City Sci. **48**, 1972–1988 (2020)
4. Degel, D., Wiesche, L., Rachuba, S., Werners, B.: Reorganizing an existing volunteer fire station network in Germany. Socio-Econ. Plann. Sci. **48**(2), 149–157 (2014)
5. http://www.mohurd.gov.cn/
6. Church, R.L., ReVelle, C.S.: Theoretical and computational links between the p-median, location set-covering, and the maximal covering location problem. Geogr. Anal. **8**(4), 406–415 (1976)
7. Yang, L., Jones, B.F., Yang, S.H.: A fuzzy multi-objective programming for optimization of fire station locations through genetic algorithms. Eur. J. Oper. Res. **181**(2), 903–915 (2007)
8. Yao, J., Zhang, X., Murray, A.T.: Location optimization of urban fire stations: access and service coverage. Comput. Environ. Urban Syst. **73**, 184–190 (2019)
9. Aktaş, E., Özaydın, Ö., Bozkaya, B., Ülengin, F., Önsel, Ş: Optimizing fire station locations for the istanbul metropolitan municipality. Interfaces **43**(3), 240–255 (2013)
10. Bolouri, S., Vafaeinejad, A., Alesheikh, A., Aghamohammadi, H.: Minimizing response time to accidents in big cities: a two ranked level model for allocating fire stations. Arab. J. Geosci. **13**(16), 1–13 (2020). https://doi.org/10.1007/s12517-020-05728-6
11. Plane, D., Hendrick, T.: Mathematical programming and the location of fire companies for the denver fire department. Oper. Res. **25**, 563–578 (1977)
12. Schilling, D., Revelle, C., Cohon, J., Elzinga, D.: Some models for fire protection locational decisions. Eur. J. Oper. Res. **5**, 1–7 (1980)

13. Badri, M., Mortagy, A., Alsayed, C.: a multi-objective model for locating fire stations. Eur. J. Oper. Res. **110**, 243–260 (1998)
14. Murray, A.: Optimising the spatial location of urban fire stations. Fire Saf. J. **62**, 64–71 (2013)
15. Sakawa, M., Kato, K., Sunada, H., Shibano, T.: Fuzzy programming for multi-objective 0–1 programming problems through revised genetic algorithms. Eur. J. Oper. Res. **97**, 149–158 (1997)
16. Tzeng, G.-H., Chen, Y.-W.: The optimal location of airport fire stations: a fuzzy multi-objective programming and revised genetic algorithm approach. Transp. Plan. Technol. **23**, 37–55 (2007)
17. Farahani, R.Z., Hekmatfar, M., Fahimnia, B., Kazemzadeh, N.: Hierarchical facility location problem: Models, classifications, techniques, and applications. Comput. Ind. Eng. **68**, 104–117 (2014)
18. Şahin, G., Süral, H., Meral, S.: Locational analysis for regionalization of Turkish Red Crescent blood services. Comput. Oper. Res. **34**(3), 692–704 (2007)
19. Gourdin, E., Labbé, M., Yaman, H.: Telecommunication and location (2001)
20. http://www.geatpy.com/
21. http://www.szft.gov.cn/
22. Chen, J., Yang, S., Li, H., Zhang, B., Lv, J.: Research on geographical environment unit division based on the method of natural breaks (Jenks). Int. Arch. Photogramm. Remote Sens. Spat. Inf. Sci. **3**, 47–50 (2013)
23. Zhao, M., Chen, Q.: Risk-based optimization of emergency rescue facilities locations for large-scale environmental accidents to improve urban public safety. Nat. Hazards **75**(1), 163–189 (2014). https://doi.org/10.1007/s11069-014-1313-2
24. Deb, K., Pratap, A., Agarwal, S., Meyarivan, T.: A fast and elitist multi-objective genetic algorithm: NSGA-II. IEEE Trans. Evol. Comput. **6**(2), 182–197 (2002)
25. Konstantinidis, A., Yang, K., Zhang, Q.: An evolutionary algorithm to a multi-objective deployment and power assignment problem in wireless sensor networks. In: IEEE GLOBECOM 2008-2008 IEEE Global Telecommunications Conference. IEEE (2008)
26. Men, J., et al.: A multi-objective emergency rescue facilities location model for catastrophic interlocking chemical accidents in Chemical Parks. IEEE Trans. Intell. Transp. Syst. **21**(11), 4749–4761 (2019)
27. Audet, C., Bigeon, J., Cartier, D., Le Digabel, S., Salomon, L.: Performance indicators in multiobjective optimizatio. Eur. J. Oper. Res. **292**(2), 397–422 (2020)
28. Sun, Y., Yen, G.G., Yi, Z.: IGD indicator-based evolutionary algorithm for many-objective optimization problems. IEEE Trans. Evol. Comput. **23**(2), 173–187 (2018)
29. Fonseca, C.M., Paquete, L., López-Ibánez, M.: An improved dimension-sweep algorithm for the hypervolume indicator. In: 2006 IEEE International Conference on Evolutionary Computation. IEEE (2006)

Utilization of Machine Learning Algorithm to Determine Factors Affecting Response to Action Among Filipinos Toward the Eruption of Taal Volcano

Ardvin Kester S. Ong[1], Yogi Tri Prasetyo[1,2(✉)], Yoshiki B. Kurata[1,3], and Thanatorn Chuenyindee[4]

[1] Mapúa University, Metro Manila, Philippines
{aksong,ytprasetyo}@mapua.edu.ph, ybkurata@ust.edu.ph
[2] Yuan Ze University, 135 Yuan-Tung Rd., Zhongli 32003, Taiwan
[3] University of Santo Tomas, España Blvd., 1015 Manila, Philippines
[4] Navaminda Kasatriyadhiraj Royal Air Force Academy, Bangkok 10220, Thailand
thanatorn_chu@rtaf.mi.th

Abstract. The Taal volcano in the Philippines has been declared to erupt soon, rising to level 3 of 5 levels. The purpose of this study was to determine factors affecting response to action among Filipinos with regards to the eruption of Taal volcano using machine learning algorithm. Collecting 501 respondents through convenience sampling online, decision tree and random forest classifier resulted to consider eruption and evacuation characteristics (EV), disaster experience, and assessment damage (AD) led to the positive response to action (AD). Further utilization of artificial neural network showed that AD and EV were main factors that greatly affected RA. The result discussed that knowledge on how the damage may affect livelihood and infrastructure, together with experience would lead to high levels of RA, mitigation, and to follow government officials. This study contributes greatly to natural disaster preparation and mitigation. Moreover, the model considered could greatly impact the evaluation of factors to determine mitigation of other natural disasters as well.

Keywords: Natural disaster · Taal volcano · Machine learning algorithm · Artificial neural network

1 Introduction

Out of 15 most dangerous volcanoes, Taal being one of which from the Philippines is currently at level 3 on the 5-tier scale [1, 2]. The raise of the Taal volcano danger poses a great threat with the people as it recently spewed a 1-km high plumes of steam and gas. Delos Reyes et al. [3] ranked the Taal volcano as the second most active out of 24 active volcanos in the country. Taal volcano had a total of 34 eruptions since the first one in 1572 [4]. The mitigation of the people should be highlighted and measured by

F. Neri et al. (Eds.): CCCE 2022, CCIS 1630, pp. 181–192, 2022.
https://doi.org/10.1007/978-3-031-17422-3_17

measuring their response to action since the authorities advised an evacuation of the people in proximity last July 8, 2021.

The behavior of people with mitigation and action plans have been dealt with. Recent studies measured these using descriptive statistics and even a powerful statistical analysis tool such as structural equation modeling (SEM) [4–7]. The measurement of mitigation and plan of action has been measured by different studies, however, the current rise on the danger of the explosion of Taal volcano has been underexplored. Moreover, machine learning algorithm has also become a trend for predicting human behavior. Specifically, Machine Learning Algorithms (MLA) such as Decision Tree (DT) and Artificial Neural Network (ANN) has been utilized to predict behaviors, mitigation, and prediction of future events [8–17].

Yang and Zhou [18] described DT as an MLA that classifies datasets to measure factors such as human behavior. In addition, Milani et al. [19] promoted the use of DT for determining factors affecting human behavior. On the other hand, ANN is an MLA that specializes on recognizing and predicting a target output. Kimes et al. [20] stated that ANN is a powerful tool that can measure and predict complex relationships of different factors. Therefore, different studies indicated that DT and ANN could be utilized for classifying and characterizing human factors such as mitigation and plan of action.

In the Philippines, a study conducted by Ong et al. [4] measured the mitigation of Filipinos with an earthquake as natural disaster. The study focused on utilizing structural equation modeling which indicated that knowledge of the natural disaster leads to a positive mitigation among people. Prasetyo et al. [5] studied about the response action of Filipinos for the Taal volcano. The study utilized SEM for the prediction of factors. Recent studies have focused more on the recognition and perception of hazards and risks [6]. Gaillard [7] studied about volcanic risk perception but focused on Mt. Pinatubo in the Philippines. With the constant natural disasters happening in the Philippines, Venable et al. [21] focused only on typhoon risk perception among Filipinos. With that, there has not been many studies that dealt with the Taal volcano in the Philippines. Specifically, focusing on response action among Filipinos using Machine Learning Algorithm (MLA).

Despite the numerous studies that dealt with MLA for natural disasters, there has been no studies focusing on the Taal volcano. Burry et al. [8] utilized artificial neural network (ANN) on ecosystem modeling. The study claimed that ANN could be a tool for measuring past events such as volcanic eruptions. Shin et al. [9] also utilized ANN for measuring multi-hazard assessment considering mitigation for building frames. Their study discussed how MLA can be an advantage by rapidly processing and performing large number of data for analysis. Seht [10] utilized ANN for the detection of seismic signal of volcanoes in Indonesia. The overview of the signal could be well established using the ANN.

The numerous studies utilizing ANN as MLA showed promising result for predicting behaviors for natural disasters. However, there has been scares information about mitigation and plan of action about volcanic eruption using ANN. In addition, Kim et al. [11] utilized MLA such as DT and DT-Random Forest Classified (DTRFC) to measure seismic facies. Results showed that DTRFC increased the accuracy in predicting the seismic facies and the determination of reservoirs. In addition, Snehil and Goel [12] utilized DT for natural disaster such as floods. Lastly, Chen et al. [13] utilized DTRFC and ANN for evaluating risk in China regarding flood disaster.

With that, the aim of this study was to predict factors affecting response to action among Filipinos with the Taal volcano. This study considered factors such as eruption characteristics, disaster experience, asset damage, socio-demographic characteristics, and evacuation characteristics. Figure 1 presents the proposed framework utilized in this study. This is the first study that utilized MLA to determine and characterize factors affecting plan to action among Filipinos with the Taal volcano. By the utilization of DT, DTRFC, and ANN, the results of this study will be beneficial among government officials to help mitigate the citizen during natural disasters. Moreover, the determination of factors would help people in mitigation and reduction of perceived severity from natural disasters. The findings of this study could be applied and extended to other natural disasters, not just volcanoes around the world.

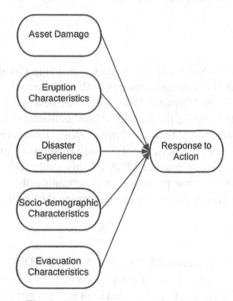

Fig. 1. Propose theoretical framework.

2 Methodology

2.1 Participants

A total of 501 participants who voluntarily answered a self-assessed online questionnaire were considered in this study. All participants were Filipinos collected via convenience sampling method. Table 1 presents the demographics of this study. There were 55.4% male and 44.6% female wherein 54.2% are within the age range of 15–24, 20.9% are 25–34 years old, 13.7% are 35–44 years old, and 11.2% were older than 34. The province considered in this study were areas near the Taal volcano, 81.9% coming from Batangas and 16.3% coming from Manila, 1.7% represents the farther area.

Table 1. Descriptive statistics of the respondents.

Characteristics	Category	N	%
Gender	Male	278	55.4
	Female	224	44.6
Age	15–24	272	54.2
	25–34	105	20.9
	35–44	69	13.7
	45-older	56	11.2
Province	Batangas (near)	411	81.9
	Manila (mid)	82	16.3
	Pampanga (far)	8	1.6

2.2 Questionnaire

The collection of data was through an online survey utilizing Google Forms. A 5-point Likert scale was utilized to obtain the measures of the respondents, wherein 5 measured "Strongly Agree" and 1 for "Strongly Disagree". The demographics as presented in Table 1 as the first section, while the second section considered Eruption Characteristics (G), Assessment Damage (AD), Socio-demographic characteristics (S), Evacuation Characteristics (EV), Disaster Experience (EX), and Response Action (RA) (Table 2).

Python 3.8 was utilized to run the MLA of this study. DT, DTRFC, and ANN underwent different optimization processes in determining the best parameters. The subsections explain the processes involved to determine the optimum result.

2.3 Questionnaire

DT was utilized as a classification MLA tool in this study. Different parameters were set such as the Gini Index (Eq. 1) and Entropy Index (Eq. 2) following different studies [14, 15]. Moreover, Milani et al. [15] stated that DTs are best utilized as a nonparametric classification method since it does not need any assumptions.

$$Gini(t) = 1 - \sum_j \left[p(j|t)\right]^2 \tag{1}$$

$$Entropy(t) = - \sum_j p(j|t) \log p(j|t) \tag{2}$$

It was seen from the optimization the Gini Index, compared to Entropy Index with Best splitter compared to Random splitter with Max Depth set to 4 had the higher accuracy rate. Figure 2 presents the DT result of this study. With 100 runs each, 70:30 training:testing ratio got the highest accuracy rate. However, the optimum tree only got an average of 59% among the different training:testing ratios considered (90:10, 80:20, and 70:30). Following the suggestion of other studies [11–13], DT with Random Forest Classifier may be utilized to enhance the average accuracy rate.

Table 2. Constructs of the study.

Items	Questions	Supporting References
G1	Due to my location, I have experienced volcanic eruption for months	Riede [23]
G2	Due to my location, I have experienced an explosive eruption	Riede [23]
G3	Due to my location, I have experienced volcanic eruption frequently	Riede [23]
G4	I know how far I am from the volcano	Riede [23]
G5	I am fully aware that I have minimal time to prepare from the volcanic eruption	Riede [23]
AD1	I have seen damaged roads after the volcanic eruption	Pan et al. [24]
AD2	I have seen destroyed buildings after the volcanic eruption	Pan et al. [24]
AD3	I have experienced serious detrimental effects on crops and livestock after the volcanic eruption	Pan et al. [24]
AD4	I have experienced severe home damage after the volcanic eruption	Pan et al. [24]
AD5	I have experienced ecological damage after the volcanic eruption	Huhtamaa et al. [25]
AD6	I have experienced transportation problems after the volcanic eruption	Van Manen [26]
S1	I have experienced a disruption of school classes after a volcanic eruption	Doocy et al. [27]
S2	I have experienced subsequent disruption of social networks after a volcanic eruption	Riede [23]
S3	I have considered moving to another place to work or to live	Riede [23]
S4	I observed that the local government is hands-on in handling the calamity	Doocy et al. [27]
S5	I have experienced a disruption of income generation after a volcanic eruption	Doocy et al. [27]
EV1	There is an evacuation center in the shortest distance from home	Witham [28]
EV2	There is immediately provided time for officials to activate emergency response whenever there is a volcanic eruption	Witham [28]
EV3	Traffic enforcers are handling and regulating the flow of traffic amidst the situation	Witham [28]
EV4	There are enough services provided during the evacuation as a result of political intervention	Andreastuti et al. [29]

(continued)

Table 2. (*continued*)

Items	Questions	Supporting References
EV5	I am willing to leave the town whenever there is an explosive eruption	
EX1	I experienced a volcanic eruption before that is why I am prepared and have the knowledge for this situation	Van Manen [26]
EX2	I experienced an earthquake before that is why I am prepared and have the knowledge for this situation	
EX3	I experienced a typhoon before that is why I am prepared and have the knowledge for this situation	
RA1	There is an increased number of tourists observed after the volcanic eruption	Van Manen [26]
RA2	There is a drastic negative change in the environment after the volcanic eruption	Van Manen [26]
RA3	There is a decrease in economic movement after the volcanic eruption	Riede [23]
RA4	I got exposed to volcanic ash after the volcanic eruption that increased my respiratory symptoms such as extreme cough, or phlegm, irritated nose, etc	Van Manen [26]

2.4 Questionnaire

Utilizing Gini Index, Best splitter, Max Depth of 4, and 70:30 training: testing ratio, the result of the DTRFC after 100 runs had an average accuracy rate of 96% with 0.00 standard deviation. The results for both trees are consistent wherein factors AD, EV, and G were factors that determined RA. Figure 3 presents the optimum DTRFC for this study. Following the study of Chen et al. [13], DTRFC and ANN was utilized to evaluate risks. Thus, the next section discusses the ANN methodology.

2.5 Questionnaire

Different parameters were considered in running the ANN. Adopting from different studies [18, 26–28], sigmoid or Relu [18] activation function for hidden layer and tanh for the output layer were considered. Adam was considered the best optimizer among different runs made [9] compared to SGD [27] and RMSProp [28]. Optimization was done with 120 epochs, running 25 times each combination on the initial run was done [10]. It was seen that the highest average accuracy rate of 65.72% for sigmoid activation function of only 1 hidden layer, tanh for the output layer, and Adam as the optimizer. Only 12 nodes in the hidden layer and 10 nodes in the output layer had no overfitting.

Further optimization was done to determine the best training and testing set, considering 60:40, 70:30, 80:20, and 90:10. From the result of 150 epochs and 50 runs each, the 80:20 training: testing ratio had the optimum result with accuracy rate of 71.30%

for AD factor followed by 69.78% for EV. Demonstrated in Fig. 4 is the ANN model utilized in this study. The input layer had 5 nodes, representing the 5 factors considered in this study in predicting RA. Moreover, Fig. 5 shows the optimum training and testing output.

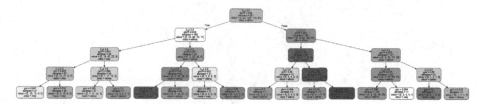

Fig. 2. Optimum decision tree.

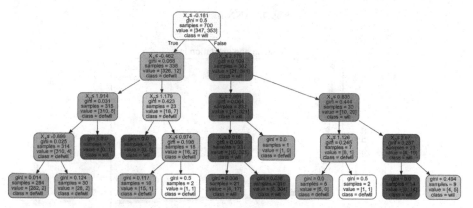

Fig. 3. Optimum decision tree with random forest classifier.

3 Results and Discussion

3.1 Decision Tree

From Fig. 2, it could be seen that G played a factor in decision with RA. Whether having value less or greater than or equal to 3.5, it would lead to the S factor considering less than or equal to 3.5. Considering the same parameters, EX will be considered leading to AD if it was satisfied. This leads to a decision for RA to "may prepare". If it was not satisfied, it will lead to same factor however having less than or equal to 4.5 that would lead to "will not" have RA. If the second child, S was satisfied, it will consider G, otherwise EX. Either factor would lead to "will not" have RA unless EX had a value less than or equal to 3.5, then people "may" have RA.

If G was not satisfied, it will lead to S with values less than or equal to 3.5. If this was satisfied, then it leads to AD with less than or equal to 3.5. Either decision would lead to EV and EX with "will not" have RA. If S was satisfied, it leads to S with value less than

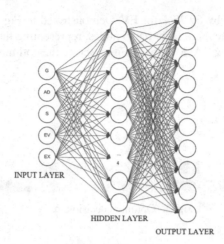

Fig. 4. ANN model.

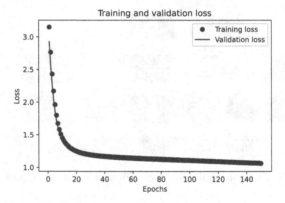

Fig. 5. Optimum training and testing result.

or equal to 4.5. If satisfied, it will consider EX with value less than or equal to 4.5. If satisfied, people "may" have RA, otherwise "will not" have RA. If S was not satisfied, then it will lead to AD considering less than or equal to 3.5. If this was satisfied, people "may" have RA, otherwise "definitely not" have RA.

It could be interpreted that G played a crucial role with RA. Considering the constructs of the study, people are aware of the risk and the severity of the eruption of Taal volcano. However, due to lack of experience and negative side-effects, they do not perceive much danger as seen in EX. This led to less mitigation among people near the area that may be affected with the explosion of Taal volcano.

3.2 Decision Tree with Random Forest Classifier

Based from the results, it could be seen that AD played a role for people to have RA. With negative perception of AD (≥ -0.181), it will lead to higher negative perception of AD (≥ -0.462). Either satisfied or not, it will lead to G. This will eventually lead to "definitely will" have RA if AD and S are considered.

If AD was not satisfied, it will consider G with value less than or equal to 0.109. If this was satisfied, it will consider EV, leading to AD, otherwise people "definitely will" have RA. AD having a value of less than or equal to 0.018 would lead to "will" have RA. If the 2nd child was not satisfied, then it will lead to AD with value less than or equal to 0.835. Satisfying this will lead to EV with value less than or equal to 1.126 that will cause people to "definitely will" have RA. Otherwise, G will be considered leading to "will" have RA.

Based from the constructs of the study, if people have experienced how highly impactful the volcanic eruption negatively causes, this will lead to high response action for mitigation and evacuation. Moreover, the destruction of infrastructures and high damage of facilities would lead people to have highly positive response action. Mainly the negative experience would lead people away from the possible impact of natural disasters such as earthquake.

3.3 Artificial Neural Network

Based from the ANN result, the 2 factors that played a huge role in RA were AD and EV with 71.30% and 69.78% average accuracy rate, respectively. The results are in line with the findings of the other MLA tools. Based from the constructs of AD, the more destruction, negative impact, and damage the volcanic eruption causes, the more people will have RA. They will be willing to leave everything to be safe due to the natural calamities that may be occurring or high probability occurring. Moving to evacuation centers and listening to public officials were considered under EV. The experience or possible knowledge of the impact caused by the volcanic eruption would give urgency among people to prepare and mitigate during natural calamities.

3.4 Artificial Neural Network

In the Philippines, the Taal volcano erupted 2 months before the COVID-19 pandemic. It was considered as one of the most severe disaster that happened in the early 2020. Asset Damage (AD) was the highest factor that affected Response Action among Filipinos. With high number of respondents were residing close to the area, the damage was clearly evident as factor that affected their decision. The livelihood as a socio-economic factor together with infrastructures and accessibilities to information affected the knowledge and impact of natural disasters such as volcanic eruptions. This is in line with the results of van Manen [22] that discussed how socio-economic factors affected people's perception of mitigation.

Based from the results, EV played a crucial role in people's mitigation. Riede [19] had a result similar to this study about risk response. The damage of houses and roads led people to move towards evacuation centers, leaving everything behind. Aside from that, a positive response towards government officials were evident during the incident. Moreover, Andreastuti et al. [25] discussed how preparedness within a community can be effective if people in the community have the ability to perceive the risks, hazards, and early-warning signs. Thus, knowledge about natural disasters and mitigation should be incorporated for people to actively respond to natural disasters especially people living near prone areas.

Despite the impact and evident results of this study, there were some limitations that still need to be considered. The study considered only 5 factors that could possibly affect the response action among Filipinos. Other factors such as measurement of knowledge, preparation/preparedness, plans, and the culture of the community could be variables that may affect RA. The low accuracy rate of ANN is also considered as a limitation. ANN is utilized to for pattern recognition and prediction. The higher accuracy would serve as a vital tool to recognize the patterns. Therefore, researchers may consider other parameters upon optimization to possibly increase the accuracy rate. Lastly, the study focused more on people near the area. It would be interesting to compare results of different distances that may be affected by the volcanic eruption. There may be findings that can be vital towards creating a mitigation plan in volcanic eruption and other natural disasters.

4 Conclusion

The Taal volcano in the Philippines is presumed to erupt soon. The status of the volcano has rose from level 2 to level 3 in a 5-level scale during the early month of July 2021. The aim of this study was to determine the factors affecting Filipino's response to action with the eruption of Taal volcano. Utilizing machine learning algorithm, specifically decision tree, decision tree with random forest classifier, and artificial neural network, results showed that the highest factor affecting Filipino's response to action was AD (71.30%) and EV (69.78%).

People will generally have high response to action if they have enough knowledge of the natural disaster. In addition, if people had the experience of the negative impact towards livelihood, infrastructure, and other socio-economic factor, then they will have high level of response to action. Moreover, people will generally follow the governments advise and policies to mitigate and prepare towards natural calamities such as volcanic eruption. It was seen that DT, DTRFC, and ANN could be utilized to predict factors that determines people's behavior towards natural disaster.

References

1. Reuters, Philippines raises Taal volcano danger level as thousands evacuate, CNN, (2021). https://edition.cnn.com/2021/07/01/asia/mass-evacuation-philippines-volcano-intl-hnk/index.html
2. Zlotnicki, J., et al.: Combined electromagnetic geochemical and thermal surveys of Taal Volcano (Philippines) during the period 2005–2006. Bull. Volcanol. **71**, 29–47 (2009). https://doi.org/10.1007/s00445-008-0205-2
3. Delos Reyes, P.J., Bornas, M.A.V., Dominey-Howes, D., Pidlaoan, A.C., Magill, C.R., Solidum, R.U.: A synthesis and review of historical eruptions at Taal Volcano, Southern Luzon, Philippines. Earth Sci. Rev. **177**, 565–588 (2018). https://doi.org/10.1016/j.earscirev.2017.11.014

4. Ong, A.K.S., et al.: Factors affecting intention to prepare for mitigation of "the bing one" earthquake in the Philippines: integrating protection motivation theory and extended theory of planned behaviour. Int. J. Disaster Risk Reduction **63**, 102467 (2021). in press
5. Prasetyo, Y.T., et al.: Factors affecting response actions of the 2020 Taal Volcano eruption among Filipinos in Luzon, Philippines: a structural equation modeling approach. Int. J. Disaster Risk Reduction **63**, 102454 (2021). in press
6. Williams, L., Arguillas, F., Arguillas, M.: Major storms, rising tides, and wet feet: adapting to flood risk in the Philippines. Int. J. Disaster Risk Reduction **50**, 101810 (2020). https://doi.org/10.1016/j.ijdrr.2020.101810
7. Gaillard, J.: Alternative paradigms of volcanic risk perception: the case of Mt. Pinatubo in the Philippines. J. Volcanol. Geoth. Res. **172**(3–4), 315–328 (2008). https://doi.org/10.1016/j.jvolgeores.2007.12.036
8. Burry, L.S., Marconetto, B., Somoza, M., Palacio, P., Trivi, M., D'Antoni, H.: Ecosystem modeling using artificial neural networks: an archaeological tool. J. Archeological Sci.: Rep. **18**, 739–746 (2018). https://doi.org/10.1016/j.jasrep.2017.07.013
9. Shin, J., Scott, D.W., Stewart, L.K., Jeon, J.S.: Multi-hazard assessment and mitigation for seismically-deficient RC building frames using artificial neural network models. Eng. Struct. **207**, 110204 (2020). https://doi.org/10.1016/j.engstruct.2020.110204
10. Ibs-von Seht, M.: Detection and identification of seismic signals recorded at Krakatau volcano (Indonesia) using artificial neural networks. J. Volcanol. Geoth. Res. **176**(4), 448–456 (2008). https://doi.org/10.1016/j.jvolgeores.2008.04.015
11. Kim, Y., Hardisty, R., Torres, E., Marfurt, K.J.: Seismic facies classification using random forest algorithm. In: SEG Technical Program Expanded Abstracts, pp. 2161–2165 (2018). https://doi.org/10.1190/segam2018-2998553.1
12. Snehil, R.G.: Flood damage analysis using machine learning techniques. Procedia Comput. Sci. **173**, 78–85 (2020). https://doi.org/10.1016/j.procs.2020.06.011
13. Chen, J., Li, Q., Wang, H., Deng, M.: A machine learning ensemble approach based on random forest and radial basis function neural network for risk evaluation of regional flood disaster: a case study of the Yangtze River Delta, China. Int. J. Environ. Res. Public Health **17**(49), 1–21 (2019). https://doi.org/10.3390/ijerph17010049
14. Yang, W., Zhou, S.: Using decision tree analysis to identify the determinants of residents' CO2 emissions from different types of trips: a case study of Guangzhou, China. J. Clean. Prod. **277**, 124071 (2020). https://doi.org/10.1016/j.jclepro.2020.124071
15. Milani, L., Grumi, S., Camisasca, E., Miragoli, S., Traficante, D., Di Blasio, P.: Familial risk and protective factors affecting CPS professionals' child removal decision: a decision tree analysis study. Child Youth Serv. Rev. **109**, 104687 (2020). https://doi.org/10.1016/j.childyouth.2019.104687
16. Kimes, D.S., Nelson, R.F., Manry, M.T., Fung, A.K.: Review article: attributes of neural networks for extracting continuous vegetation variables from optical and radar measurements. Int. J. Remote Sens. **19**(14), 2639–2663 (2010). https://doi.org/10.1080/014311698214433
17. Venable, C., Javernick-Will, A., Liel, A.B., Koschmann, M.A.: Revealing (mis)alignments between household perceptions and engineering assessments of post-disaster housing safety in typhoons. Int. J. Disaster Risk Reduction **53**, 101976 (2021). https://doi.org/10.1016/j.ijdrr.2020.101976
18. Canario, J.P., de Mello, R.F., Curilem, M., Huenupan, F., Rios, R.A.: Llaima volcano dataset: in-depth comparison of deep artificial neural network architecture on seismic events classification. Data Brief **30**, 105627 (2020). https://doi.org/10.1016/j.dib.2020.105627
19. Riede, F.: Doing palaeo-social volcanology: developing a framework for systematically investigating the impacts of past volcanic eruptions on human societies using archaeological datasets. Quatern. Int. **499**(10), 266–277 (2019). https://doi.org/10.1016/j.quaint.2018.01.027

20. Pan, H., Shi, P., Ye, T., Xu, W., Wang, J.: Mapping the expected annual fatality risk of the volcano on a global scale. Int. J. Disaster Risk Reduction **13**, 52–60 (2019). https://doi.org/10.1016/j.ijdrr.2015.03.004

21. Huhtamaa, H., Helama, S.: Distant impact: tropical volcanic eruptions and climate-driven agricultural crises in seventeenth-century Ostrobothnia, Finland. J. Hist. Geogr. **57**, 40–51 (2017). https://doi.org/10.1016/j.jhg.2017.05.011

22. van Manen, S.M.: Hazard and risk perception at Turrialba volcano (Costa Rica); implications for disaster risk management. Appl. Geogr. **50**, 63–73 (2014). https://doi.org/10.1016/j.apgeog.2014.02.004

23. Doocy, S., Daniels, A., Dooling, S., Gorokhovich, Y.: The human impact of volcanoes: a historical review of events 1900–2009 and systematic literature review. PLoS Currents (2013). https://doi.org/10.1371/currents.dis.841859091a706efebf8a30f4ed7a1901

24. Witham, C.S.: Volcanic disasters and incidents: a new database. J. Volcanol. Geoth. Res. **148**(3–4), 191–233 (2005). https://doi.org/10.1016/j.jvolgeores.2005.04.017

25. Andreastuti, S., Paripurno, E.T., Gunawan, H., Budianto, A., Syahbana, D., Pallister, J.: Character of community response to volcanic crises at Sinabung and Kelud volcanoes. J. Volcanol. Geoth. Res. **382**, 298–310 (2019). https://doi.org/10.1016/j.jvolgeores.2017.01.022

26. Simek, C.K., Arabaci, D.: Simulation of the climatic changes around the coastal land reclamation areas using artifical neural networks. Urban Clim. **38**, 100914 (2021). https://doi.org/10.1016/j.uclim.2021.100914

27. Jena, R., et al.: Integrated model for earthquake risk assessment using neural network and analytical heirarchy process: Aech province Indonesia. Geosci. Frontiers **11**, 613–634 (2020). https://doi.org/10.1016/j.gsf.2019.07.006

28. Yousefzadaeh, M., Hosseini, S.A., Farnaghi, M.: Spatiotemporally explicit earthquake prediction using deep neural network. Soil Dyn. Earthq. Eng. **144**, 106663 (2021). https://doi.org/10.1016/j.soildyn.2021.106663

Author Index

Printed in the United States
by Baker & Taylor Publisher Services